Advances in Surface Engineering
Volume II: Process Technology

Advances in Surface Engineering
Volume II: Process Technology

Edited by

P.K. Datta
University of Northumbria at Newcastle, UK

J.S. Burnell-Gray
University of Northumbria at Newcastle, UK

THE ROYAL
SOCIETY OF
CHEMISTRY
Information
Services

Sep/ae Chem (handwritten)

The Proceedings of the Fourth International Conference on Advances in Surface Engineering held at The University of Northumbria at Newcastle on 14–17 May 1996.

The front cover illustration is taken from the contribution by S.K. Ibrahim, A. Watson and D.T. Gawne, p.245.

Special Publication No. 207

ISBN 0-85404-752-2

A catalogue record for this book is available from the British Library

Published by The Royal Society of Chemistry,
Thomas Graham House, Science Park, Milton Road,
Cambridge CB4 4WF, UK

Printed and bound by
Bookcraft (Bath) Ltd

SD 12/16/97 (handwritten)

Preface

Advances in Surface Engineering is based on the Proceedings of the *4th International Conference on Advances in Surface Engineering* which was hosted by the University of Northumbria's *Surface Engineering Research Group* between 14–17th May 1996.

Acknowledgements

The editors wish to thank Professor Gilbert Smith, the Vice Chancellor of the University of Northumbria for opening the conference. The editors are grateful to Professor Van de Voorde for giving the opening technical keynote talk.

The editors wish to express their gratitude for the support extended by:

The Department of Trade and Industry, The Institute of Materials, The Institute of Corrosion, The Royal Society of Chemistry, Multi-Arc (UK) Ltd, Buehler Ltd, Gearing/Micromaterials, Mats (UK), Tech Vac and Woodhead Publishing.

Special thanks are due to Professor Cryan head of the *School of Engineering* at the *University of Northumbria at Newcastle*.

The support and encouragement of many colleagues at the *University of Northumbria at Newcastle* and friends in other universities, is gratefully acknowledged.

The human commitment to any conference or book is substantial and often not fully acknowledged. In this regard the work of Kath Hynes, the secretaries and the technicians from the *School of Engineering* and the members of the *Surface Engineering Research Group (SERG)* should be fully recognized.

Finally special commendation is reserved for Dan Smith who administered the *4th International Conference on Advances in Surface Engineering (4ICASE)* and David Griffin of *SERG* who desk-top published the conference proceedings.

P. K. Datta
J. S. Burnell-Gray

Surface Engineering Research Group
School of Engineering
University of Northumbria at Newcastle

Contents
Volume II Process Technology

Section 2.5 Electrochemical and Electroless

Contents
Volume I Fundamentals of Coatings

Introduction

Section 1.1 High Temperature Corrosion

Section 1.2 Aqueous Corrosion

Section 1.4 Fatigue and Other Failure

Contents
Volume III Engineering Applications

Introduction

Section 3.1 Biomedical

Section 3.2 Aerospace

Introduction

J. S. Burnell-Gray and P. K. Datta

SURFACE ENGINEERING RESEARCH GROUP, UNIVERSITY OF NORTHUMBRIA AT
NEWCASTLE, UK

1 SCOPE OF *ADVANCES IN SURFACE ENGINEERING*

Advances in Surface Engineering is aimed at reviewing and documenting the recent advances in research and application of this relatively newly emerging technology, the problems that remain to be solved and the directions of future research and development in surface engineering.

The three volumes incorporate both science and technical research papers. They demonstrate how SE technologies are continuously increasing the level of performance of components, devices and structures through the creation of high performance surfaces.

What distinguishes *Advances in Surface Engineering* is an attempt, though limited, to characterise the coatings/engineered surfaces in terms of their fundamental structural entities and to understand their behaviour and properties using the principles of material science and physics. This knowledge of coating structures together with an understanding of the mechanisms of degradation processes that operate on surfaces[1-3], has allowed the development of precisely designed surfaces/coatings with enhanced degrees of corrosion resistance, wear resistance and biocompatibility.

More broadly, *Advances in Surface Engineering* provides a lens for viewing fundamental changes in the SE and corrosion and wear management professions. In this era of advanced manufacturing technologies and virtual networks, most factors of production are available globally. Capital essentially flows freely; machines can be bought or their capacity rented; technology and technological mastery are readily transferable. What increasingly sets research, and its application apart are knowledge and expertise – *intellectual*, as opposed to physical, assets.

New and more powerful ways to think about problems and actions are the prized output from the *4th International Conference on Advances in Surface Engineering*. New theories enable engineers to conceptualize their activities in novel ways and initiate more effective programmes of action. New theoretical inspiration – ranging across a broad spectrum – can also help academic researchers redefine their efforts. Particularly prized is a new methodology or way of thinking that totally transforms the shape of a field and the leveraged efforts of hundreds of researchers.

Advances in Surface Engineering is structured in 3 volumes. **Volume 1** concerns fundamental aspects of corrosion and wear. **Volume 2** gives an appreciation of SE technologies. **Volume 3** deals with applications of surface engineering to selected industrial sectors – areas central to surface engineering and holding particular promise for improvements in existing and emerging surface engineering techniques.

The *Introduction* to *Advances in Surface Engineering* aims to provoke new debate and comprehension to devise a coherent and integrated framework for tackling the enduring engineering problem of understanding and controlling corrosion and wear exploiting SE as a sustained business asset. Embedded in the *Introduction* are attempts to identify:

1. What were the important issues in SE in 1992 (the time of the *3rd International Conference on Advances in Surface Engineering*)?
2. What has happened in SE since 1992?
3. What are the important SE issues in 1997?
4. What are likely to be the important issues – particularly relating to management – in SE in 2010?

Finally, we take the opportunity to present a synopsis of the activities of the University of Northumbria's (UNN) *Surface Engineering Research Group (SERG)*.

2 SURFACE ENGINEERING

2.1 Background

The growing importance of surface engineering is due to the realization that modified, treated and coated surfaces can prevent degradation processes more effectively, particularly those which originate at surfaces[4]. This applies to a wide range of engineering applications, as exemplified by the wear coatings listed in Table 1.

It is now widely recognized – see for instance the applications cited in Table 1 – that the successful exploitation of these processes and coatings may enable the use of simpler, cheaper and more easily available substrate materials, with substantial reduction in costs, minimization of demands for strategic materials and improvement in fabricability and performance. In demanding situations where the technology becomes constrained by surface-related requirements, the use of specially developed coating systems may represent the only real possibility for exploitation[4,6–9].

Table 1 *Industries and components using thermally applied wear coatings[5]*

Aero gas turbines	Land-based turbines	Others
* Turbine and compressor blades, vanes	• Turbine and compressor buckets, vanes, nozzles	♦ Feed rolls
* Gas path seals	• Piston rings (IC engine)	♦ Pump sleeves
* Mid-span stiffeners	• Hydroelectric valves	♦ Shaft sleeves
* Z-notch tip shroud	• Boiler tubes	♦ Gate valves, seats
* Combustor and nozzle assemblies	• Wear rings	♦ Rolling element bearings
* Blade dovetails	• Gas path seals	♦ Dies and moulds
* Flap and slat tracks	• Impeller shafts	♦ Diesel engine cylinder
* Compressor stators	• Impeller pump housings	♦ Hip joint prostheses
		♦ Hydraulic press sleeves
		♦ Grinding hammers
		♦ Agricultural knives

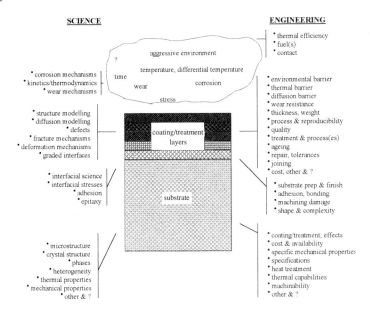

Figure 1 *Aspects of surface engineering research[1]*

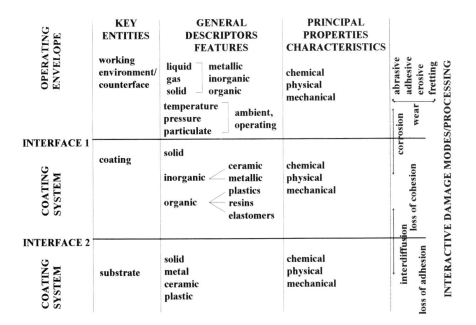

Figure 2 *Generalised features of a working coating system[6]*

Surface engineering produces surfaces with a unique combination of bulk and surface properties resulting in the creation of a high performance composite material. However, the biggest benefit that flows from the use of surface engineering lies in the ability to create new surfaces with highly non-equilibrium structures.

2.2 Corrosion- and Wear-Related Failures

The basic features of a simple wear-or corrosion-resistant coating system are shown in Figures 1 and 2. Refering to Figure 2, selective interaction is required at *Interface 1* to provide a wear- or corrosion-resistant surface. Selective interaction of a different kind is required at *Interface 2* to obtain adequate adhesion. Such interaction(s) must not lead to the removal of coating constituents and/or their dilution by interdiffusion across *Interface 2*. The requirement for prolonged sustainability of the corrosion- and wear-resistance of the coatings imposes additional constraints on the design of the coating system. The coatings must contain a reservoir of elements to sustain the required selective interactions at the surface (*Interface 1*). Other constraints on the coating design may flow from the necessity of a surface to resist a number of degradation processes occurring at the same time. To satisfy these requirements at both surfaces or to prevent conjoint actions of different modes of degradation, multicomponent/multiphase and multilayered coatings are required. Even so, adequate coating systems can now be designed and produced using intelligent combinations of various processes[10,11].

2.3 Surface Engineering Technologies

Surface engineering techniques generally consist of surface treatments where the compositions/structures or the mechanical properties of the existing surface are modified, or a different material is deposited to create a new surface. Surface engineering is essential in the application and exploitation of high performance engineering components. This is especially true in relation to both the rising costs of advanced performance structural materials and the increasingly high life-cycle costs associated with high performance systems. Table 2 illustrates the market share for various types of surface finish[12].

Deposition procedures, include traditional electrodeposition and chemical conversion coating, together with thermal spraying – where a plasma or electric arc melts a powder or wire source, and droplets of molten material are sprayed on to the surface to produce a coating; PVD, in which a vapour flux is generated by evaporation, sputtering or laser ablation; and CVD, where reaction of the vapour phase species with the substrate surface produces a coating.

Surface treatments include:
* mechanical processes that work-harden the surface – e.g. shot-peening;
* thermal treatments which harden the surface by quenching constituents in solid solution – e.g. laser or electron beam heating;
* diffusion treatments which modify the surface composition – e.g. carburizing and nitriding;
* chemical treatments that remove material or change the composition by chemical reactions – e.g. etching and oxidation; and
* ion implantation – see Table 3 for applications – where the surface composition is modified by accelerating ions to high energies and implanting them in the near-surface[4].

Table 4 lists characteristics of a coating which are important in relation to quality assurance.

Table 2 *Surface finishing – value by industry sectors*[12]

Coating	Size (£M)	Share (%)
Organic	1,450	43
Plating	705	21
Galvanizing	355	10
Surface heat treatment	325	10
Hard facing	100	3
Anodizing	65	2
Tin plating	40	1
Vitreous enamelling	40	1
PVD & CVD	25	1
Others	250	8
Total	3,355	100

Comparisons of certain of the above techniques are contained within Table 5 and examples of deposition and treatment technologies used in the aerospace industry are given in Table 6.

2.4 State of the Art and Future Developments

Since the early 1980s there has been a continuing and rapid development of advanced surface engineering practices for the optimization of corrosion and wear resistance. It is now possible to produce coatings of novel composition and microstructure in multilayer/ multicomponent format as appropriate to the design specification, by a variety of sophisticated physical and chemical processes, including hybrid technologies. The paradox concerning compatibility between the environment, coating and substrate is no longer a problem. At the level of research scientists and engineers, efforts must be made to enhance and systematize understanding of the various process technologies. For instance in PVD, one such issue – which offers a distinct competitive advantage – concerns plasma densities and their importance

Table 3 *Industrial exploitation of ion implantation*[13]

Material	Application (specific examples)	Typical results
Cemented WC	Drilling (printed circuit board, dental burrs etc)	Four times normal life, less frequent breakage and better end product
Ti–6Al–4V	Orthopaedic implants (artificial hip and knee joints)	Significant (400 times) lifetime increase in laboratory tests
M50, 52100 steel	Bearings (precision bearings for aircraft)	Improved protection against corrosion, sliding wear and rolling contact fatigue
Various alloys	Extrusion (spinnerets, nozzles and dies)	Four to six times normal performance
D2 steel	Punching and stamping (pellet punches for nuclear fuel, scoring dies for cans)	Three to five times normal life

Table 4 *Characteristic properties of a coating*

Structural	Mechanical	Physical	Chemical
Composition	Adhesion	Specific heat	Chemical stability
Density	Cohesion	Thermal expansion	Environmental compatibility
Porosity	Hardness		Corrosion resistance
Phase contents	Modulus		Biocompatibility
Crystallinity	Ductility		
Grain size	Strength		
Amorphosity	Fracture toughness		
Defect structures	Internal stress		
Dislocations	Wear resistance		
Vacancies	Friction coefficient		
	Deformation mode		

in facilitating the repeatable manufacture of advanced surface engineered artefacts with outstanding properties and performances[10,11]. However, a further critical problem with regard to quality assessment is the precise significance of measured properties and characteristics – e.g. hardness – in relation to actual coating performance. Here insight is needed into the consequence of particular hardness levels in relation to wear performance, so that a coating engineered to a given hardness could be expected to offer a predetermined design wear life.

At the fundamental level there is a need to understand structure/property relationships in, for example coatings, so that surface engineered systems can be designed from conception to develop desired properties. This also requires a better understanding of the degradation processes which need to be prevented/minimized by the designed surface. In this regard there is considerable scope for the creation of tailored coatings of chosen composition, structure and properties – including multilayer/multicomponent format – by highly adaptable PVD and CVD technologies (Table 7).

Table 5 *Comparison between five surface engineering processes[14]*

Process	Resistance to wear	Risk of distortion	Resistance to impact	Convenience	Range of materials
Plasma spraying – atmospheric	High	Low	Low	Very good, gun is offered to the work	Extensive
Plating	High	Low	Medium	Low, work is processed in a bath	Low
Welding	Medium	High	Good	Good	Medium
CVD/PVD ion deposition	High	Low	Good	Low vacuum chamber required	Good
Cladding	Low	Low	Good	Good	Low

Table 6 *Surface engineering technologies used in the aerospace industry[15]*

Technique	Material	Requirement	Application
Mechanical treatments, eg peening	Steels, titanium-based and nickel-based alloys	Improved mechanical and wear properties	Compressor blade roots
Paints	Phenolic and epoxy polyurethanes	Cosmetic, corrosion and wear, earthing, emissivity and infrared	Shafts, discs, blading
Polishing	Steels, titanium-based and nickel-based alloys	Cosmetic, salvage and repair efficiency	Aerofoil surfaces on vanes and blades
Electrochemical	Tribomet, chromium	Corrosion and wear, salvage and repair	Bearing chambers, stator vanes
Thermal spraying (D-gun, flame spraying, plasma spraying)	Al/Si polyester, WC/Co, CuNiIn	Corrosion and wear, salvage and repair, seals, net-shapes	Snubbers, gas-path seals, combustor cans
Thermochemical	Nitrogen and carbon into steels	Improved mechanical properties	Shafts and gears
Pack aluminizing	Nickel-based alloys	Corrosion/oxidation	Aerofoil surfaces on vanes and blades

2.5 Quality Assurance

Quality assurance of surface engineered coatings and surfaces is a major issue particularly for the coating users and producers[10].

In the absence of a definitive knowledge of the structure/property relationships in coatings deposited on a surface, only an empirical approach can be adopted[10]. A coating can be described in terms of its characteristic properties (Table 4).

One approach which is being increasingly adopted is to define the functionality of the coating for a particular application in terms of a sub-set of the properties listed above. For example, to achieve quality assurance of a load-bearing prosthesis, consideration can be given, in the first instance, to the sub-set of properties consisting of adhesion, strength, wear resistance, friction and biocompatibility. Similarly corrosion resistant coatings can be quality assured by addressing parameters such as adhesion, residual stress, ductility, K_{Ic}, fatigue/crack growth resistance and chemical stability.

Quality assurance of deposition technology[17], as well as manufactured and surface engineered artefacts to reliably impart specified measurable performances, is central to surface engineering. Effort needs to be made to develop expert systems to design and select coating/treatment systems and hence define the appropriate process technology.

These skills and competencies must be applied within the construct of business realities – achieving sustained world-class competitive advantage.

Table 7 *Future surface engineering activities*[4,16]

1. Surface engineering of non-ferrous metals
2. Surface engineering of polymers and composites
3. Surface engineering of ceramics
4. Mathematical modelling of surface engineered components
5. Surface engineering in material manufacture
6. Statistical process control in surface engineering
7. Non-destructive evaluation of surface engineered components
8. Duplex or hybrid surface engineering technologies and design, eg:
   * laser treatment of thermal and plasma spray coatings
   * ion beam mixing and ion-assisted coatings
   * hot isostatic pressing of overlay coatings
   * thermochemical treatment of pre-carburized steels
   * thermochemical treatment of pre-laser hardened steels
   * CVD treatment of pre-carburized steels
   * PVD treatment of pre-nitrided steels
   * ion implantation of pre-nitrided steels

3 MANAGEMENT ISSUES

Hard and anecdotal findings from research into the strategic management of technology and the exploitation of technological innovation suggest that, for leading firms in a wide variety of industries, developing advanced technologies *per se* is rarely the constraining challenge in technological innovation. Rather, the challenge is innovation in the market. In this regard quality management, management of change, strategic management, innovation and knowledge management are as important as technological issues.

3.1 Quality

Bench-marking is a means of measuring and comparing performance and may be defined as, "the continuous process of measuring products, services and practices against the toughest competitors or those companies recognized as industry leaders". The emphasis should be on understanding how surface technologists carry out their activities; learning how other groups excel in surface engineering; and then adapting and reinterpreting what has been learnt in a way that makes for competitive advantage. Instead of aiming to improve only against previous performance, technologists should use bench-marking to inject an element of imagination into the quest for progress, while simultaneously and objectively scrutinizing established processes. Inevitably, the perspective is international[18,19].

As part of the normal research work an external audit could be carried out, this may comprise a survey of the controls surrounding researchers' daily records – before and after implementation of quality improvements. Assessments of the following areas could be among the audit's key functions:
* assurance of daily task completion;
* periodic review by management;
* verifiability of results claimed against physical evidence;
* evidence of consistency/quality of results claimed; and
* indication of appropriate coordination with clients.

Items for further investigation might also include:
- How is control exercised by management – formally or informally?
- Is control proportional to risk, exposure and objectives?
- Are guidelines for the delegation of authority suitable for the organization?

A standard part of the audit would be an analysis of the effectiveness of researchers' time management. The *day in the life of* technique consists of unobtrusively observing the activities of selected researchers over a period of several days. The allotment of their – *value added* or *non-value added* – time should be reviewed with both the researcher and manager. For *non-value added* activities the root causes of the operation would be investigated and minimized[20].

3.2 Change

Management of strategic and organizational change must address three key questions:
1. How is change initiated and implemented in relatively successful organizations?
2. What is the rôle of the management at the service provider and end-user companies in initiating change?
3. What is the contribution of management development in the implementation of organizational and strategic change?

Change management must at the same time stimulate innovation and provide mechanisms for dealing with uncertainty in knowledge-intensive SE business environments. This might be accomplished by systematically analyzing the organization's values, needs, interests and relationships, and productively applying the insights gained. An important aim of this process is to determine – using multiple levels of comparison – specific factors that might lead to competitive advantage.

Individual organizations and consortia need to assess complementary plant networks – here the objective should be to provide qualitative insight and analytical tools to facilitate the development and adaptation of plant networks in response to diverse and changing markets, manufacturing costs and technological capabilities across countries[21].

3.3 Strategy

During times of tumult strategic alliances become increasingly attractive. Research has indicated that to make strategic alliances work, success does not necessarily come from the structural or systems aspects of the alliance, but rather from the quality of relationships, the degree of trust, mutual commitment and the flexibility of attitude brought to the relationship. General assessments of the management challenges associated with creating and managing strategic alliances must consider the dynamics of cross-functional partnerships and a specific evaluation of the rôle of international alliances in high-tech SE industries. Alliances need to be analyzed from three perspectives:
a. direct economic costs and benefits;
b. historical evolution; and
c. external networks.

Growing reliance on strategic alliances has prompted investigations into how alliance partners utilize complementary technological capabilities to create products that neither could develop individually and what drives companies to choose strategic alliances and joint ventures over internal R&D and licensing as approaches to augmenting technological capabilities. Also being explored are how technology-based companies develop and exploit strategically valuable knowledge assets and the managerial and technical processes by which novel technologies are applied to the development of complex new products. Other studies are suggesting ways in which senior executives can nurture superior product development performance and advancing the notion that technological innovation is often less a constraint than the need to innovate in the marketplace. Likewise, research suggests the principal challenge in entrepreneurship is building skills such as selling ideas and products and applying theories of competitive advantage, not communicating new knowledge or developing new theories that better explain entrepreneurship. Of particular interest is how technological knowledge is transferred across company boundaries and the rôle collaborative development might play in countering a company's core rigidities, i.e. deeply in-grained, but out-dated, technological capabilities.

Strategic decision-making may be considered as essentially a technique for making judgements when the outcome depends in part on the actions of others, it involves systematic analysis consistent with the commonplace expedient of putting oneself in another's shoes, and interactive decision and value analysis. The process may involve clinical, statistical and theoretical research focused on major commitments, notably investments and disinvestments.

The impact of technology on industry structure – such as investments in SE technologies – need to be evaluated as part of coherent business strategies and be viewed as strategic necessities rather than attempts to gain sustainable advantage. In considering technological change and competitive strategy it is necessary to explore the dynamic links between and strategic consequences of technological change and shifts in organizational structure and competitive advantage. Research attempts to answer questions such as:

* How does a firm identify opportunities to create value from technological change and substitution?
* How can firms create sustainable profits in the face of SE technological change and resulting adjustments in competitive dynamics?
* What methods are available to managers for evaluating highly uncertain projects and the value of developing specific competencies and capabilities?
* How might competitors' capabilities be assessed?
* How does technological change affect competitive dynamics and redefine industry structures?
* Under what conditions should a firm invest in a new technology?
* How do competitive dynamics affect the evolution of technology and attendant quality standards?

The merging of computers, telecommunications and SE – and the blurring of functional and technological boundaries within the SE industry – forces managers to pursue a wide range of topics, including the competitive dynamics within and between quality standards, the rôle of alliances in SE technology, and the relationship between technology choices, scope of a firm (i.e. degree of horizontal and vertical integration) and financial decisions.

Research might profitably be conducted into the determinants of superior SE process development performance and process development strategies within, for instance, the aerospace and automotive industries. Detailed qualitative and quantitative data could be collected on the histories, strategies and performance of surface engineered artefacts. Statistical analysis could then be used to identify how such factors as organizational structure, project strategy, organizational capabilities and technological environment influence process development lead times, productivity and costs. Such a study would be expected to shed light on the special challenges that attend the management of R&D projects and building of development capabilities in SE-based industries; in addition this would yield significant in-sights into the potential strategic rôle of SE technologies in these important industries over the coming decade[21].

3.4 Innovation

Innovation – the successful exploitation of new ideas – is the key to sustained competitiveness. British university research groups have responded to the Government's "Science, Engineering and Technology" and "Competitiveness" White Papers, and especially that aspect which is concerned with the application of new techniques and ways of working that improve the effectiveness of individuals and organizations. The scope of several research programmes covers the rôle of innovative management in the achievement of sustained improvement in the bottom-line performance of commercial and industrial businesses. In particular emphasis is placed on the human and organizational processes and conditions that contribute to this. Such "research on business in business" is designed to increase knowledge and understanding of these crucial elements and to encourage their exploitation by industry[22].

Organizations frequently manage innovation and development activities not only as single projects but also as a cohesive set of related projects. Research is currently underway in a number of organizations to compare the different methods used for managing project sets. The aim is to identify "best practice" methods and to examine the link between the strategic aims of the organization and the formation of the goals and objectives of the projects[23].

3.5 Knowledge

In new lean, business processes where non-value added tasks have been eliminated, IT (information technology) can facilitate manufacturing philosophies, e.g. JIT (just in time). However, complex problem areas still exist in re-engineering companies. These problems will occur with large and small companies, and basic IT cannot address them. They include the handling of incomplete, conflicting and vague data, the discovery of knowledge in massive data sets, the interpretation of legislation and inter-organizational contracts, the management of change, and the re-application of an expert's accrued experience and expertise. *Knowledge-based systems* (KBSs) provide a series of techniques that can help to assess, manage and ameliorate these problems. There have been a number of reported successes where KBSs have added value to business processes, and in fact made business re-engineering possible[24].

A *knowledge-based system* refers to any assemblage which incorporates a level of expertise and experience, which can be used to address new situations in an intelligent way. The knowledge may derive from human *experts*, research papers and reports, or computer systems. KBS technology – e.g. expert systems, neural networks, data mining and artificial intelligence – applied to SE can offer industry significant benefits:

- intelligent decision-making support, making safer, faster and more effective decisions;
- consistency of approach and assurance of quality standards;
- better service delivery, increased productivity and improved cost control;
- dissemination of scarce expertise across the organization;
- a valuable training tool for new engineers and managers; and
- developing a way of making complex situations more transparent for the decision-maker and linking shop-floor activity to commercial transactions.

The technical focus might be concentrated in the fields of knowledge-based SE and corrosion and wear management, and it is in these areas that developed solutions to complex industrial problems should be sought[25].

4 THE *SURFACE ENGINEERING RESEARCH GROUP*

4.1 Background

The *Surface Engineering Research Group* (*SERG*) comprises a Director (Prof. P. K. Datta), an Assistant Director (Dr J. S. Burnell-Gray), 5 academic consultants, 2 Research Fellows, 2 Research Associates and 4 Research Students. It has five core functions: *Research, Education, Technology Transfer, Consultancy* and *Training*.

SERG firmly believes that surface engineering is one of the several keys to UK industry gaining a world-class competitive advantage. The *Group's* objectives are to identify outstanding pivotal research issues, promote the take-up of SE research results in industry and influence UK and European Union research policy to help create the next generation of manufacturing systems.

Our technical focus is concentrated in the fields of surface engineering, corrosion and wear, and it is in these areas that we seek to develop commercial applications as solutions to complex industrial problems. The research focus of the laboratory is addressed along with partners in for instance Rolls-Royce, Multi-Arc (UK), Chromalloy UK and Johnson Matthey. *SERG's* research portfolio also supports cross-fertilization of the work of other groups.

The development of successful applications in corrosion and wear management is not a straightforward task as it requires the synergistic combination of expertise on coatings deposition, surface engineering, corrosion and wear engineering, as well as corrosion/wear theory. However, the rewards of SE are across-the-board improvements in value, quality, customer support and productivity.

4.2 Research Portfolio

It is within this demanding and continually changing framework – also see Figure 3 – that the *Group's* research activities are based. Added-value to the component in relation to enhanced mechanical, thermal, chemical, electrical or optoelectrical attributes, as well as fitness-for-purpose and the minimization of life cycle costs, are the driving forces for the justification and adoption of surface engineering practices. In this regard the *Group's* research has not only contributed to the characterization and understanding of the functional behaviour of coatings, but also at a more fundamental level has involved the systematic application and testing of scientific principles with a view to creating entirely new types of surfaces with novel properties. More pragmatically this research effort has in addition contributed to the achievement of more reliable coatings with reproducible properties. This reproducibility aids prediction of the performance, e.g. mechanisms and time-dependent behaviour, of surface-modified artefacts using surface analytical techniques, system modelling, interfacial simulation and NDE – all of which are essential if engineers are to fully exploit the potentials of the discipline.

The scope of research programmes – from basic, through strategic/pre-competitive to applied – is outlined in Table 8. Other examples of applied surface engineering research lie not only in the field of aero engine turbine blades, but also in natural gas production and combustion, and orthopaedics.

The *Surface Engineering Research Group* advocates the principles of surface engineering and provides local and national industry with a world-class facility for the study of corrosion and wear control using surface engineering technologies. Since the early 1980s *SERG* has gained an international reputation in the area of surface engineering and also acts as a regional teaching facility in corrosion and wear prevention using the latest surface engineering technologies involving surface analysis, surface modification and coatings deposition. *SERG* offers a well-founded mechanical engineering workshop and corrosion laboratory comprising computer-aided machining, non-traditional machining, and high temperature gaseous and molten salt, as well as aqueous corrosion facilities.

Table 8 *SERG's research portfolio*

Current and Recent Programmes

- *Design and Optimization of High Temperature (HT) Protective Coatings* concerns the design and development of HT degradation resistant MCrAlYX-type coating systems capable of withstanding corrosion in coal gasifier atmospheres typically containing significant Cl_2 or S_2 potentials. The project formed part of the EPSRC Rolling Programme in Surface Engineering jointly pursued by ourselves, Hull University and Sheffield Hallam University.
- *Interfacial Modification of MMCs* aiming to improve HT mechanical properties and corrosion resistance of Ti-Ti alloy matrix/SiC fibres. Interfacial modelling is used to quantitatively and qualitatively describe diffusion and corrosion mechanisms. HT studies are performed in atmospheres designed to simulate aero-engine compressors. Part of the EPSRC Rolling Programme with UNN support.
- *Optimization of Electroless Deposition Processes for Protective Coatings* studying and exploiting selected parameters (bath chemistry, and coating composition and morphology) of electroless Ni-B and Ni-P coatings with varying concentrations of B and P, as a function of coating condition and determining the resultant effect on corrosion, wear and fatigue behaviour. Part of the EPSRC Rolling Programme.
- *Design and Development of Pt-Aluminide Coatings for Improved Corrosion Resistance* determining a reference database and enhancing Pt-modified aluminide coatings, deposited with or without Ta, on superalloy substrates designed for use as gas turbine blades. Hot corrosion is monitored in simulated gas turbine environments. Sponsored by EPSRC, Rolls-Royce plc, C-UK Ltd. and Johnson Matthey.
- *Improved High Temperature Resistant Silicon Nitride-Silicon Carbide Composites* concerning the characterization and optimization of Si_3N_4-SiC composites during exposure to replicated combustion environments. The EU provided funds for this research programme, jointly pursued by ourselves, Limerick University, T&N Technology and British Gas.
- *High Temperature Corrosion of Car Engine Valve and Valve Seat Materials* involving the study and selection of appropriate Cr_2O_3- and Al_2O_3-forming alloys, oxide dispersion-containing and other mechanically alloyed materials, to optimize durability in simulated car engine atmospheres. Funding by EPSRC and British Gas.
- *Surface Engineered (Diamond-Like Carbon) Prostheses* embraces a wide range of activities, viz: depositing and characterizing DLC coatings, biocompatibility tests, and fatigue and wear studies of coated prostheses. Funding is from DTI Link. Partners include 3M, Teer Coatings and the Royal Victoria Infirmary (Newcastle).
- *Studies of Electroless Coatings* concerning the deposition and characterization of electroless coatings on a number of substrates. The aim was to produce coatings with a set of specific properties such as high corrosion resistance, texture, appearance and good adhesion. The project was sponsored by NPL.

Completed Programmes

* *Coatings Technology*
  - Development of Ta and Nb ion-plated coatings as novel load-bearing (human) implant materials

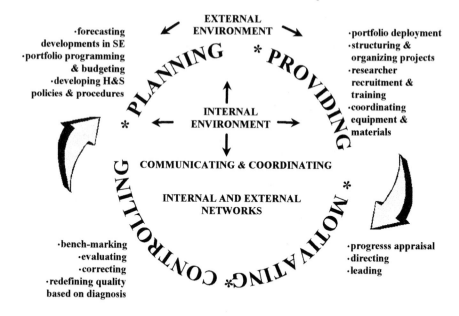

Figure 3 *SERG's research management functions*

- Electrolytic deposition of Co-Sn alloys for the electrical industry
* *Aqueous Corrosion*
  - Corrosion of implant systems in bodily fluids
  - Corrosion of electroless coating systems
  - Passivation in electrolytically deposited alloys
* *Environmental Cracking*
  - Stress corrosion cracking (SCC) in titanium alloys used in submarine hulls
  - Development of software and hardware to study and analyse crack propagation
  - SCC of nodular cast irons in heavy goods vehicle (HGV) suspensions
* *High Temperature Degradation*
  - Chloridation of binary alloys, MCrAlX-type and MCrAlY-type coatings alloys
  - Sulphidation/oxidation of advanced engineering ceramics, Si_3N_4/SiC composites, CoCrAlYX-type coatings alloys, refractory metals, HfN and Nb_2N
  - Oxidation of titanium alloys and MMCs for aerospace applications

---

4.3 Future Challenges, Development and Plans

Plans to consolidate and expand on the success enjoyed by *SERG* centre on:

Research
- ◆ Expanding into interfacial modelling and interfacial engineering;
- ◆ moving into the area of pack cementation;
- ◆ diversifying into deposition (PVD) control and optimization;
- ◆ assessing the viability of PVD catalyse;

◆ moving into new pre-treatments, e.g. pre-nitridation and pre-carburization;

◆ extending electroless deposition into cobalt alloys;

◆ moving away from sulphidation towards chloridation and erosion/corrosion;

◆ introducing the concept of knowledge-based systems;

◆ contingency funding to refurbish/maintain existing equipment;

◆ buying a mass spectrometer;

◆ constructing a burner rig; and

◆ further exploiting IPR.

Education

◆ Offering the facility of split PhDs, interchange of research fellows and travelling professors;

◆ providing a surface engineering module to existing postgraduate and EPSRC IGDS courses; and

◆ developing an IGDS in Research Management.

Technology Transfer

◆ Presenting the 5th International Conference on Advances in Surface Engineering in 2000;

◆ hosting the 1998 Institute of Corrosion's Corrosion Science Symposium;

◆ hosting the 2nd European Workshop in Surface Engineering Technologies for SMEs in 1998;

◆ promoting in-house courses in corrosion, wear and surface engineering based on EPSRC's Rolling Programme in Surface Engineering commencing 1998; and

◆ further enhancing consultancy and contractual services to industry.

References

1. J. S. Burnell-Gray and P. K. Datta (editors), 'Surface Engineering Casebook' Woodhead Publishing, November 1996.

2. K. N. Strafford, P. K. Datta and J. S. Gray (editors), 'Surface Engineering Practice: Processes, Fundamentals and Applications in Corrosion and Wear', Published by Ellis Horwood, 1990.

3. P. K. Datta and J. S. Gray (editors), Conf Proc 3rd Int Conf Advances in Surface Engineering, Newcastle upon Tyne, May 1992. Published by The Royal Society of Chemistry, 1993, 'Surface Engineering Vol I: Fundamentals of Coatings', 'Surface Engineering Vol II: Engineering Applications', 'Surface Engineering Vol III: Process Technology and Surface Analysis'.

4. V. Sankaran, 'Surface Engineering – A Consultancy Report', Advances in Materials Technology: Monitor, Issue 24/25, February 1992.

5. P. Sahoo, *Powder Metallurgy International*, 1993, **25**, 73.

6. K. N. Strafford and S. Subramanian, *J. Materials Processing Technology*, 1995, **53**, 393.

7. Reference 2, Keynote paper, Chapter 3.1.1, p. 397.

8. J. S. Burnell-Gray and P. K. Datta (editors), 'Quality Issues in Surface Engineering', To be published by Woodhead Publishing 1997.

9. J. S. Burnell-Gray, Internal reports, MBA course, University of Sunderland , 1995-1997.

10. Correspondence with K. N. Strafford, University of South Australia.
11. Correspondence with K. N. Strafford, University of South Australia.
12. D. Hemsley, *Engineering*, 1994, **235**, 25.
13. P. Sioshansi, *Thin Solid Films*, 1984, **118**, 61.
14. I. H. Hoff, *Welding and Metal Fabrication*, 1995, **63**, 266.
15. D. S. Rickerby and A. Matthews, 'Advanced Surface Coatings: A Handbook of Surface Engineering', Published by Blackie, 1991.
16. T. Bell, *J. Phys D: Appl Phys*, 1992, **25**, A297.
17. C. Subramanian, K. N. Strafford, T. P. Wilks, L. P. Ward and W. McMillan, *Surface and Coatings Technology*, 1993, **62**, 529.
18. A. van de Vliet, *Management Today*, January 1996, 56.
19. F. C. Allan, *Special Libraries*, 1993, **84**, 123.
20. S. J. Burns, *Internal Auditor*, 1991, **48**, 56.
21. Anon, Harvard Business School, Internet home pages: www.hbs.edu/research/summaries/lec.html, accessed 21st October 1996.
22. Anon, Economic & Social Research Council, Internet home pages: www.bus.ed.ac.uk:8080/ESRC-Innovation.html, accessed 23rd October 1996.
23. Anon, Cranfield University, School of Management, Internet home pages: www.cranfield.ac.uk/som/res/default.html, accessed 18th October 1996.
24. Anon, Ulster University, Northern Ireland Knowledge Engineering Laboratory, Internet home pages: www.nikel.infj.ulst.ac.uk/nintro.htm, accessed 23rd October 1996.
25. Anon, London University, London Business School, Internet home pages: http://www.lbs.lon.ac.uk/om/research.html#tech: accessed 18th October 1996.

Section 2.1 PVD and CVD

2.1.1
Recent Developments in Magnetron Sputtering Systems

R. D. Arnell and P. J. Kelly

RESEARCH INSTITUTE FOR DESIGN, MANUFACTURE, AND MARKETING, THE UNIVERSITY OF SALFORD, SALFORD, UK

1 INTRODUCTION

Surface engineering techniques now play a vital rôle in many industrial sectors, such as manufacturing, automotive, aerospace, nuclear, chemical, and power generation. Within these industries, the most widespread applications of surface engineering techniques are, currently, in the fields of wear resistance and corrosion protection. The use of surface engineering techniques allows the engineer to specify independent surface and bulk properties for a component, and corrosion resistant coatings are routinely applied to a wide range of components. The coatings serve not only to extend service life, but also to avoid the costs incurred in producing entire components from exotic corrosion-resistant alloys. The need to improve performance, whilst reducing costs, has led to the acceptance throughout industry of the concept of surface engineering as an integral part of component design.

The Surface Engineering Laboratory at Salford University has concentrated, over the last twenty years, on the development of physical vapour deposition (PVD) processes for the deposition of a wide range of high quality and novel surface coatings. Of these processes, unbalanced magnetron sputtering, an extension of the basic sputtering process, has been developed to the stage where it can be routinely used to deposit very high quality, well-adhered coatings of a wide range of metals and ceramics, able to meet the most stringent requirements of many current engineering applications.

In the following sections of this paper, the basic sputtering process and the development and applications of unbalanced magnetron sputtering are described. Brief descriptions of three current projects are also given to illustrate the versatility of the technique.

2 THE BASIC SPUTTERING PROCESS

In its simplest form a sputtering system consists of a vacuum chamber containing a target and a substrate holder. The target (a plate of the material to be deposited) is connected to a negative DC, or RF voltage supply. The substrate holder faces the target and may be grounded, floating, or biased. It may also be heated, or cooled. Such a system is shown schematically in Figure 1. The chamber is evacuated and partially backfilled, typically to a pressure in the range 10^{-3} to 10^{-2} mbar, with a sputtering gas, usually argon, to provide a medium in which a glow discharge, or plasma may be initiated and maintained.

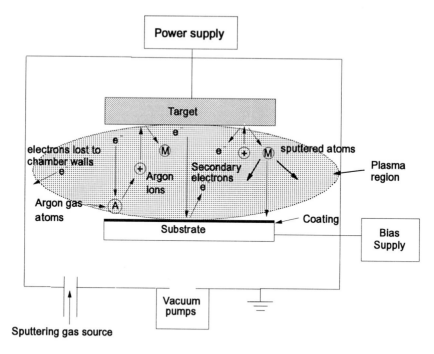

Figure 1 *Schematic representation of basic sputtering system and process*

The plasma is initiated by applying a negative voltage of the order of 2–3 kV to the target plate. Once initiated, positive ions from the plasma strike the target and eject target atoms by a momentum exchange mechanism. The ejection of target atoms in this manner is termed 'sputtering'. Some of the sputtered atoms condense on the substrate as a film. A number of other interactions occur at the target surface, caused primarily by the impinging positive ions, including the emission of secondary electrons. This causes additional ionization of the neutral sputtering gas atoms which helps sustain the discharge. A more detailed description of the sputtering process can be found in the standard work by Vossen and Kern[1].

Although many materials have been successfully deposited by the basic sputtering process, it is limited by low deposition rates, low ionization efficiency in the plasma, and high substrate heating effects. These limitations have been overcome by the introduction of what is now termed conventional magnetron sputtering and, more recently, unbalanced magnetron sputtering.

3 UNBALANCED MAGNETRON SPUTTERING

Both conventional and unbalanced magnetrons essentially consist of a water-cooled target attached to an array of magnets, or electromagnets. The magnets are arranged in such a way that the central axis of the target forms one pole and the second pole is formed by a ring of magnets around the edge of the target. By arranging the magnets in this way, a magnetic field is applied perpendicular to the electrical field at the target. This confines electrons in the

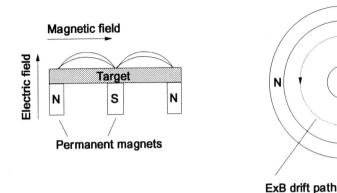

Figure 2 *A circular planar magnetron cathode, showing the magnetic field and the resulting ExB drift path*

plasma to a region near the target, resulting in increased ionization, a much, denser plasma in this region and a substantially increased film deposition rate. The basic arrangement of a magnetron is shown in Figure 2.

The increased ionization efficiency allows magnetrons to maintain a plasma at a lower operating pressure than the basic sputtering systems. At lower operating pressures there is less gas scattering of the sputtered target atoms, which further enhances arrival rates at the substrate. Thus, deposition rates can be relatively high (of the order of microns per minute). In practice the factors which control the deposition rate in magnetron sputtering are the power density in the target region, the area of the target, substrate-to-target separation, target material, and coating pressure.

The main difference between conventional and unbalanced magnetrons is in the degree to which the plasma is confined. In a conventional magnetron the plasma is strongly confined to the target region. In a typical magnetron, the region of dense plasma extends to a distance of about 6cm beyond the target. Substrates positioned inside this region are subjected to ion bombardment during film growth, which can strongly modify the structural and chemical properties of the resulting film[2]. Changes in film properties are controlled by the incident ion energy, the deposition rate and the ion current density measured at the substrate. With conventional magnetrons, due to the high degree of plasma confinement, there is a strong decrease in the substrate ion current density with *increasing substrate-to-target* separation, and substrates positioned in a very low density plasma receive insufficient ion bombardment to modify the microstructure of the growing film. It is, therefore, difficult to produce fully dense, high quality coatings on large, complex components using conventional magnetrons. This problem has been overcome by the development of the unbalanced magnetron, and its incorporation into multiple magnetron systems.

The concept of the unbalanced magnetron was first developed by Window and Savvides in 1986 when they investigated the effect of varying the magnetic configuration of an otherwise conventional magnetron[3-5]. They found that by strengthening the outer ring of magnets,

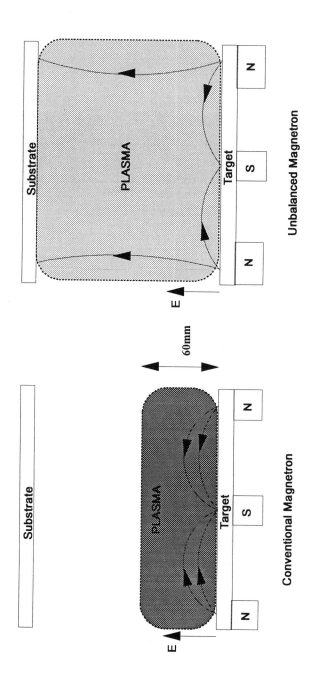

Figure 3 *A comparison of the plasma confinement in conventional and unbalanced magnetron*

electrons in the plasma were no longer confined to the target region, but were able to follow the magnetic field lines and flow out towards the substrate. By unbalancing the magnetic field in this way, a dense plasma can be confined by the magnetic field between the target and the substrate. Ion bombardment at the substrate is, therefore, increased, both because there is greater ionization near the substrate and because ions must move with the electrons to maintain the electric neutrality of the plasma. A comparison between the magnetic fields and plasma confinement in conventional and unbalanced magnetrons is given in Figure 3.

The use of unbalanced magnetron configurations, therefore, allows high ion currents to be transported to the substrate. However, it is still difficult to deposit uniform coatings onto complex components using a single magnetron source. In order to exploit this technology commercially, therefore, a number of multiple magnetron systems have been introduced. If two unbalanced magnetrons are installed vertically opposed to each other, then they can be configured with opposite magnets of the same polarity (mirrored), or of opposite polarity (closed-field). In the closed-field configuration, the magnetic field lines between the magnetrons form a closed trap for electrons in the plasma. Few electrons are therefore lost to the chamber walls and a dense plasma is maintained in the substrate region, leading to high levels of ion bombardment of the growing film. A comparison between the closed-field and mirrored configurations is shown in Figure 4. Both research and industrial scale closed-field systems have been developed by Teer, in the UK[6,7] and Sproul, in the USA[8,9]. The full name of the deposition technique is closed-field unbalanced magnetron sputtering (CFUBMS).

The technique has the capability to deposit a wide range of materials, including pure metals, alloys, multi-layers, functionally graded material and ceramics, such as oxides, nitrides and carbides. Sputtering can take place from composite targets, or powder targets, producing solid films of materials not readily available in this form. Multiple magnetron arrangements allow coatings to be uniformly deposited onto complex components, or batches of components. A wide range of ceramic materials can be deposited in a CFUBMS system by sputtering from a metallic target in the presence of a reactive gas[10–12].

Arc discharges at the target can be a problem during reactive sputtering, particularly during the deposition of highly insulating materials, such as alumina. As the coating process proceeds, areas of the targets away from the main racetrack become covered with an insulating layer. A charge builds up on this layer until breakdown, and, therefore, arcing occurs. Arc discharges at the target can lead to the ejection of droplets of material from the target, which in turn, can cause defects in the growing film. Also, the damaged area on the target can become a source of further arc discharges. This results in an increasing frequency of arcing, which prevents stable operation.

However, a new development, the pulsed magnetron sputtering process (PMS), offers the potential to overcome the problems encountered when operating in the reactive sputtering mode with the CFUBMS system. Initial studies have indicated that pulsing the magnetron discharge at medium frequencies (10–200 kHz), when depositing highly insulating materials, can stabilise the discharge. This significantly reduces the formation of arcs and, consequently reduces the number of defects in the resulting film[13–18]. Furthermore, deposition rates during pulsed reactive sputtering have been found to approach those obtained for the non-reactive sputtering of pure metal films[16,17].

If a single magnetron discharge is pulsed, then the system is described as unipolar pulsed sputtering. In this situation, the pulse-on time is limited so that the charging of the insulating layers does reach the point where breakdown, and, therefore arcing occurs. The discharge is dissipated during the pulse-off time through the plasma. If two magnetrons are connected to

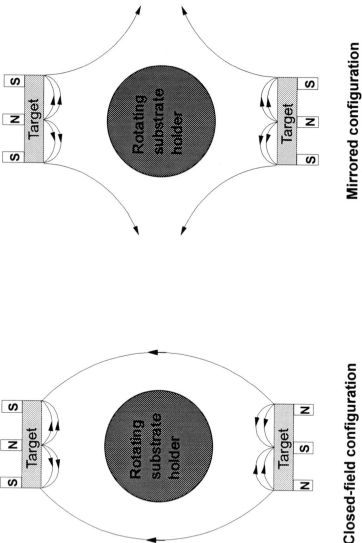

Closed-field configuration   **Mirrored configuration**

Figure 4 *A comparison of the magnetic configurations in a duel magnetron system*

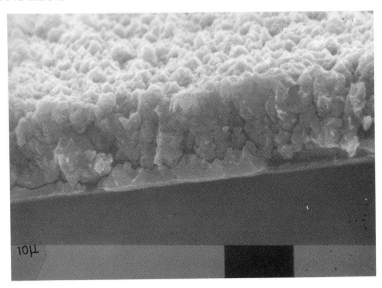

Figure 5 *Structure of an alumina film*

the same pulse supply then the configuration is described as bipolar pulsed sputtering. Each magnetron source then alternately acts as an anode and a cathode of a discharge. The periodic pole changing promotes discharge of the insulating layers, hence preventing arcing.

At Salford, the PMS process has been used to investigate the reactive sputtering of alumina films[19]. Coatings were deposited using a fixed 20kHz pulse unit. This unit, an Advanced Energy SPARC-LE unit[15], was connected in series with a standard dc magnetron driver. For

Figure 6 *SEM cros-sectional micrograph of an alumina film*

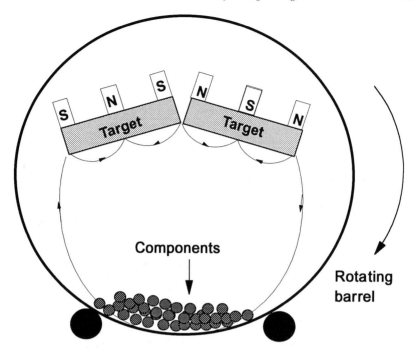

Figure 7 *Dual unbalanced magnetron barrel coater*

comparison purposes, coatings were also deposited by dc reactive sputtering without pulsing the discharge. As expected, the dc reactive sputtering of alumina films proved extremely difficult. Arcing took place from the target throughout the deposition run and the process was highly unstable. The structure of one of these coatings is shown in Figure 5. As can be seen, the coating has a granular, porous structure. It also has a sub-stoichiometric composition and a very low microhardness.

By contrast, when operating in the PMS mode with the SPARC-LE units the process was very stable, with very few arc events observed at the target. Figure 6 shows a SEM micrograph of the fracture section of an example of an alumina film deposited using the PMS process. As can be seen, the coating is extremely dense with no discernible structural aspects on the fracture surface. This coating is some 13μm thick and appears to be defect-free. The composition of this coating is close to stoichiometric Al_2O_3, and the microhardness is around 20000 MPa.

The PMS process is a major development in the reactive sputtering field. The high rate deposition of defect-free ceramic coatings onto complex components is now achievable through the use of this technique, in conjunction with the CFUBMS process.

4 EXAMPLES OF THE USE OF CFUBMS

4.1 Diamond-like Carbon

Due to their unique mechanical, physical, electrical, chemical and optical properties, diamond-like carbon (DLC) coatings are receiving more and more intense investigations. However,

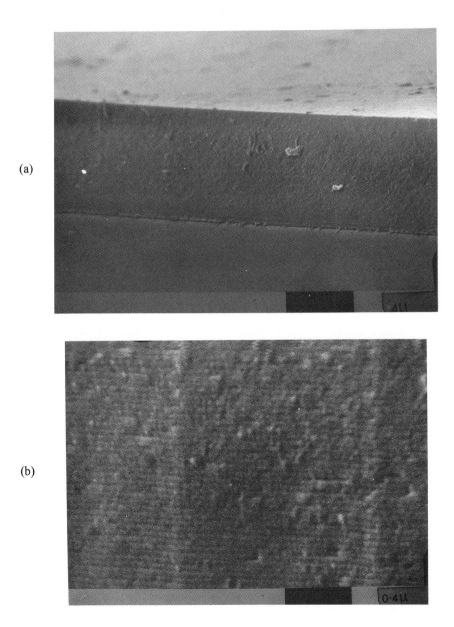

Figure 8 *(a) Low magnification SEM micrograph of fracture section of TiC multi-layer pyrotechnic coating deposited onto glass substrate by unbalanced magnetron sputtering, and (b) high magnification SEM micrograph of TiC pyrotechnic coating, revealing multi-layered structure*

until very recently, the great degree of mismatch between the properties of the coating and those of the substrate resulted in very poor coating adhesion, and effectively precluded the use of the coatings in many potentially important applications. The introduction of grading of the coatings, using the CFUBMS technique, to form multilayer structures which progressively graded the material properties from those of the substrate to those of the outer layer of the coatings, has led to great improvement in adhesion of the DLC coatings to the substrate; this has made the successful practical application of such coatings possible.

4.2 Co-Deposition of Corrosion Resistant Aluminium/Magnesium Alloys as Cadmium Replacements for the Aircraft Industry

The aircraft industry has traditionally utilised electroplated cadmium as a corrosion preventative coating on steel aircraft fitments, including fasteners. However, cadmium poses toxicity problems associated with the metal and its compounds, and, when electroplated, causes hydrogen embrittlement of high strength steel. It is for these reasons that alternatives are being sought.

For some years, the Industry has used aluminium, in the form of ion vapour deposited (IVD) coatings, for corrosion protection of steel parts. However the sacrificial properties of these films are soon lost due to the formation of oxide films on the coatings, and they have a porous, columnar structure which necessitates the application of a postdeposition glass bead peening process to consolidate the structure and render it suitable as a corrosion resistant barrier in its own right.

Recently, the University of Salford and DRA, Farnborough have developed a superior coating based on a highly-supersaturated aluminium-magnesium solid solution The coatings, of any chosen composition, are deposited using two rectangular magnetron sources, positioned side by side and pointing downwards, and with their magnetic fields opposed to create a closed field-effect. A rotating barrel is positioned to enclose the two magnetrons creating the capacity to deposit alloy films onto masses of small components (see figure 7). The components, typically airframe fasteners, nuts, screws etc., tumble randomly in the barrel within the vapour flux of the two magnetron targets, enabling each component to receive a uniform coating of the same composition and thickness.

4.3 Multi-Layer Pyrotechnic Coatings by Unbalanced Magneton Sputtering

Conventional pyrotechnic devices, such as flares, igniters, or actuators, generally consist of intimate mixtures of finely divided reactants. However, there are a number of problems associated with such devices which stem directly from the physical nature of the reactant particles. These problems include batch-to-batch variations in performance, moisture adsorption and ageing. Such problems can affect the design, manufacture and service life of a device. Another major problem is the presence of a native surface oxide layer on the fuel particles. This can inhibit reactivity and significantly influence the ignition characteristics of a device,

DRA Fort Halstead (formerly RARDE) are studying the viability of producing pyrotechnic devices using PVD techniques. This project, carried out at Salford, formed a part of this study. This is a novel application of these techniques, which offers the potential to overcome many of the problems associated with conventional pyrotechnics. At Fort Halstead, pyrotechnic coatings have been deposited by vacuum evaporation and ion plating. The materials investigated have been predominantly Mg/PTFE and various thermite-type (metal/metal oxide) compositions, and patents have been obtained to cover these applications. At Salford, unbalanced magnetron sputtering has been used to deposit titanium/carbon pyrotechnic coatings. A patent has also been obtained to cover this technique.

The titanium/carbon pyrotechnic coatings were deposited by co-deposition from a titanium target and a carbon target, i.e. sputtering took place simultaneously from both targets, whilst the substrate was rotated at a constant speed over the targets. This resulted in a very fine multi-layered structure, in which the layer thickness and the titanium-to-carbon atomic ratio can be readily controlled by varying the rotation speed of the substrate holder and by controlling the target powers. The structure of an example of one of the Ti/C pyrotechnic coatings is shown in Figures 8 (a) and (b). Using this technique, highly reactive coatings were deposited which burnt over a range of compositions from $TiC_{0.7}$ to $TiC_{1.53}$. The reactions were initiated electrically and were recorded on high speed video. Study of the videos indicated that the coatings burnt at a rate of approximately 7.5m/sec. This propagation rate is some 200 times higher than the fastest published rates for conventional Ti/C powder blends. This is a significant result which demonstrates the potential of this technique.

References

1. J. L. Vossen and W. Kern (eds), 'Thin Film Processes II'. 1991, London, Academic Press.
2. S. M. Rossnagel and J. J. Cuomo, *Vacuum*, 1988, **38**, 2, 73.
3. B. Window and N. Savvides, *J. Vac. Sci. TechnoL*, 1986, **A4**, 2, 196.
4. B. Window and N. Savvides, *J. Vac. Sci, TechnoL*, 1986, **A4**, 2, 453.
5. N. Savvides and B. Window, *J. Vac. Sci. TechnoL*, 1986, **A4**, 2, 504.
6. D. G. Teer, Proc. 7th Int'l Conf. on Ion and Plasma Assisted Techniques, IPAT '89, Geneva, Switzerland, CEP Consultants Ltd., Edinburgh, p. 145.
7. D. G. Teer, Proc. Ist Int'l Symp. on Sputtering and Plasma Processes, ISSP'91, Tokyo, 1991, p. 131.
8. A. M. Sproul, *Surf. Coat. Technol.*, 1991, **49**, 284.
9. S. L. Rohde, I. Petrov, W. D. Sproul, S. A. Bamett, P. J. Rudnik and M. E. Graham, *Thin Solid Films*, 1990, **193/194**, 117.
10. R. P. Howson, A. G. Spencer, K. Oka and R. W. Lewin, *J Vac. Sci. Technol*, 1989, **A7**, 1230.
11. Z. Pang, M. Boumerzoug, R. V. Kruzelecky, P. Mascher, J. G. Simmons and D. A. Thompson, *J. Vac. Sci. Technol.*, 1994, **A12**, 83.
12. S. Inoue, K. Tomianga, R. P. Howson, K. Kusaka, *J Vac. Sci. TechnoL*, 1995, **A13**, 2808.
13. D. A. Glocker, *J. Vac. Sci. Technol*, 1993, **A11**, 2989.
14. P. Frach, U. Heisig, C. Gottfried and H. Walde, *Surf. Coat. TechnoL*, 1993, **59**, 177.
15. W. D. Sproul, N. M. Graham, M. S. Wong, S. Lopez, D. Li and R. A. Scholl, *J Vac. Sci. Technol.*, 1995, **A13**, 1188.
16. S. Schiller, K. Goedicke, J. Reschke, V. Kirchkoff, S. Schnieder and F. Nfilde, *Surf. Coat. Technol.*, 1993, **61**, 331.
17. S. Schiller, K. Goedicke and C. Metzner. Paper presented at the Int'l Conf on Metallurgical Coatings and Thin Films (ICMCTF), April 1994, San Diego.
18. B. Stauder, F. Perry and C. Frantz, *Surf. Coat. Technol.*, 1995, **74–75**, 320.

19. P. J. Kelly, O. A. Abu-Zeid, R. D. Amell, J. Tong. Paper presented at the Int'l Conf on Metallurgical Coatings and Thin Films (ICMCTF), April 1996, San Diego, Accepted for publication.

2.1.2

Investigation of Ti$_2$N Films Deposited Using an Unbalanced Magnetron Sputtering Coating System

S. Yang, M. Ives, D.B. Lewis, J. Cawley, J.S. Brooks, and W-D. Münz

MATERIALS RESEARCH INSTITUTE, SHEFFIELD HALLAM UNIVERSITY, UK

1 INTRODUCTION

Titanium nitride is chemically stable over a broad compositional range and it is the most studied hard coating material used for improving the quality of cutting tools and extending the tool's lifetime. The structure of titanium nitride is critically dependent on its composition[1]. At high and low level of nitrogen the structure was single phase, whilst at intermediate levels of nitrogen the structure was often multiphase[2]. According to the equilibrium phase diagram[3] of TiN system, the tetragonal Ti$_2$N is an intermediate phase appearing in a very narrow concentration range (31 to 33 at % N) and exists in the temperature range below 1100°C. It is usually difficult to produce films with Ti$_2$N phase below 500°C by PVD process[4,5]. In fact, the properties of a TiN coating are critically dependent on the deposition conditions. In a PVD process, the nitrogen partial pressure must be accurately controlled[6,7]. If the nitrogen partial pressure is too low, an under stoichiometric coating will be produced which may be soft or brittle and pale in colour. However if the nitrogen partial pressure is too high, the over stoichiometric coating will be produced which normally has poor coating cohesion[8] and of course target poisoning becomes a serious problem which results in a decreased coating rate. The crystalline phases in the coating are affected by the deposition temperature[9]. The crystallographic texture and the density of the coating can be affected by the bias current density[10,11]. A higher ionisation efficiency is required in order to produce the Ti$_2$N phase[12]. The metal evaporation rate, nitrogen partial pressure, as well as the deposition temperature are known to influence the phases present during a deposition process[13]. Although many investigations have been made into the production of the Ti-N phases, it is commercially interesting to investigate the production and reproducibility of the pure Ti$_2$N film.

This paper considers the production of the Ti$_2$N films deposited using an industrial scale unbalanced magnetron sputtering coating system and describes the process parameters by which the pure Ti$_2$N was produced.

2 EXPERIMENTAL DETAILS

2.1 Coating System

The films were deposited using a HTC 1000-4 ABS system[14] which was operated in the unbalanced magnetron mode without substrate rotation. A base pressure of 1×10^{-5} mbar was achieved and the pressure during deposition was in the 10^{-3} mbar range. The substrate temperature was defined by the bias voltage, cathode power, the magnetic field used to unbalance the magnetron, and by electric heaters in the chamber. A differentially pumped mass spectrometer residual gas analyser was used to monitor the change in nitrogen gas partial pressure. The measured nitrogen partial pressure may not be the actual value in the main chamber due to the differential pumping speed in the main chamber.

2.2 Sample Preparation

The test pieces used were high speed steel and stainless steel samples prepared to a 1μm finish. The test pieces were cleaned ultrasonically and dried using hot air. A further cleaning process was performed in the chamber by heating up to 400°C and maintaining an argon ion glow discharge with a bias voltage of −1000V. The samples were subsequently etched with high energy titanium ions produced by the steered arc process in which an arc current of 100A on one target was employed with a substrate bias voltage of −1200V.

2.3 Coating Parameters

The films were deposited using only one target with a power of 10KW. A bias voltage of −120V was applied and the substrates, which were 25cm distance from the target, were coated at 480°C without rotation. The nitrogen gas flow was kept constant in one process but varied process to process from zero to a saturation level in order to give a range of nitrogen gas partial pressures by which films of different Ti:N ratio could be deposited

2.4 Analysis Instruments

Glow Discharge Optical Emission Spectroscopy (GDOES) was used to determine both the composition and distribution profile of elements within the coatings[15,16]. X-ray diffraction (XRD) was applied to determine the phase composition by comparing the diffraction peak positions with standard data[17]. Fischerscope H100 hardness and Rockwell adhesion tests were used to monitor the hardness and adhesion of the coatings. Taylor-Hobson roughness tester was used to study the roughness of the films.

3 RESULTS

3.1 Composition Dependence on the Nitrogen Partial Pressure

The variation of film composition (as measured by GDOES) with the nitrogen partial pressure (measured by a mass spectrometer differentially pumped as compared with the main chamber) is shown in Figure 1. The total chamber pressure was constant at about 2.4×10^{-3}

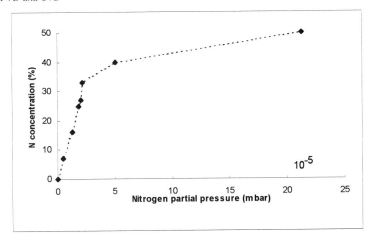

Figure 1 *Composition dependence on nitrogen partial pressure*

mbar during each process. It can be seen from this figure that there are two different stages in which the nitrogen gas partial pressure in a process affects the nitrogen concentration in the coating. The concentration of nitrogen in the films increased rapidly with a increase in nitrogen partial pressure before the reactive gas partial pressure reached approximately 2.2×10^{-5} mbar where approximately 33 at % N in the coating was produced. Beyond this point, the reactive gas partial pressure did not affect the composition as sensitively, and the concentration of nitrogen changed very slowly with the change of nitrogen partial pressure. The stoichiometric TiN (at 50% N) coating was produced at the nitrogen partial pressure range of 2.1×10^{-4} mbar.

3.2 Crystallographic Phases

The XRD patterns for films of 0, 7, and 16 at % N are shown in Figure 2. When films were produced at 0, 7 and 16 at % N, the diffraction peaks identified $TiN_{0.3}$ (as compared with the powder X-ray diffraction file data JCPDS 41-1352) which is the hexagonal (hcp) structure; however, a distinct distortion of the hcp titanium lattice was obvious as the diffraction peaks were shifted to the left indicating the solubility of nitrogen atoms in the hcp titanium lattices.

As the concentration of nitrogen in the film was increased to 25 at % N, multi-phase compositions of $\alpha TiN_{0.3}$ and εTi_2N^* (powder X-ray diffraction file JCPDS 23–1455), were found in the coating as shown in Figure 3. The $\alpha TiN_{0.3}$ (100), (002), (101), and εTi_2N^* (204), (107) were found in this film. The pure tetragonal εTi_2N film was produced in the film of 27 at % N. There were two Ti_2N phases identified (one referred to the XRD file JCPDS 17–386 and another to file JCPDS 23–1455). This is shown in Figure 3 in which almost all the peaks of the phase for the Ti_2N were identified and the strongest orientation was found to be the Ti_2N <001>. Some peaks for the second tetragonal Ti_2N^* were also identified in this sample and they were $Ti_2N^*(204)$ and (107). Both Ti_2N and TiN phases were found in the film of 33 at % N (see Figure 3). It can be seen that the Ti_2N (101), (200), (111), (311), and $Ti_2N^*(103)$, (204), (107) were identified. TiN (111) and (200) were also found in this coating.

When coatings were deposited at 40 and 50 at % N, the cubic structure of stoichiometric

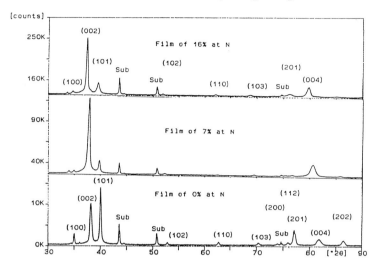

Figure 2 *XRD patterns for films of 0, 7, and 16 at % N*

titanium nitride was found. The XRD diffraction patterns for these coatings were similar to each other. The XRD pattern for the coating of 50 at % N is shown in Figure 4 in which the TiN (111), (200), (220), and (311) were identified and the strongest orientation was the <110>.

3.3 Microhardness and Rockwell Adhesion

The hardness values of these coatings as a function of nitrogen concentration are shown in Figure 5. It can be seen that the hardness linearly increases up to approximately 26000N/mm^2 as the concentration changes from 0 to 25 at % N. The hardness reduced to a value of 22000N/

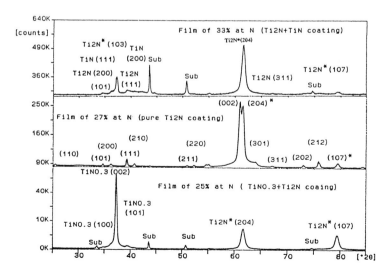

Figure 3 *XRD patterns for films of 25, 27, and 33 at % N*

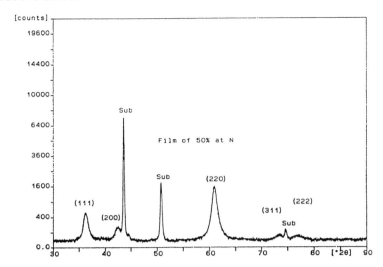

Figure 4 *XRD patterns for films of 50 at % N3.3 Micro hardness and Rockwell adhesion*

mm² when the concentration of the film (the pure Ti₂N film as measured by XRD) was 27 at % N, followed by an increasing hardness to a maximum of 32000N/mm² when the concentration of the coating increased to 40 at %. The hardness decreased to approximately 27000N/mm² when the coating contained 50 at % N. The hardness value of the pure Ti₂N coating was less than that of coatings with stoichiometric TiN phase composition as expected. However Poulek et al.[18] have conducted similar experiments and reported the result shown in Figure 5 which also fits with other reported results for the hardness of Ti₂N. In interpreting the results obtained in this work, it is speculated that the operating conditions, particularly the choice of nitrogen partial pressure, were not sufficiently discriminating to optimise the production of single phase

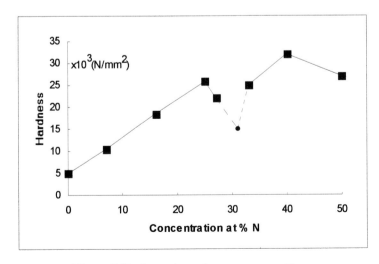

Figure 5 *Hardness dependence on composition*

- Reference(18) for the hardness of εTi₂N phase.

stoichiometric Ti_2N. Thus the dotted lines shown on Figure 5 suggest that this work could be in agreement with that of Poulek et al[18].

Excellent adhesion was found for these films as measured using the Rockwell indentation test. The indentation of $R_c = 1$ was obtained in the films with concentrations of nitrogen from 0 to 40 at % N. Good adhesion ($R_c = 2$) was also found in the coating of 50 at % N. Even though the Ti_2N film was more than 15 μm thick, it still obtained a $R_c = 1$.

3.4 Surface Roughness Dependence on Composition of the Films

Figure 6 shows the variation of roughness of the films with composition. It can be seen that the pure Ti_2N film (composition 27 at %N) had the lowest roughness (Ra = 0.11μm, Rz = 0.89 μm, Rt = 1.5 μm). The multi-phase films (αTiN + Ti_2N, or Ti_2N +TiN, composition 25 at % or 33 at %N) had the higher roughness (Ra = 0.23 or 0.20 μm, Rz = 2.41 or 3.10 μm, Rt = 3.92 or 5.00 μm respectively). The single phase films (TiN, αTiN, and pure Ti) have lower roughness compared with the multi-phase films.

4 DISCUSSION

It has been shown that the tetragonal Ti_2N phase was located in a very narrow composition range between 31 and 33 at % N. The N-rich and N-poor Ti_2N coexisted with δTiN and αTi(N) respectively. In this paper, the concentration of the film for pure Ti_2N phase was 27 at %N. This discrepancy may result from the quantitative inaccuracy of the GDOES technique. The problem associated with producing pure Ti_2N coating was to obtain the right concentration at a certain temperature and titanium sputtering rate in a critical nitrogen partial pressure regime during the coating process. Here the temperature of the deposition process can be precisely controlled since a heater was used in the main chamber, and the sputtering rate was

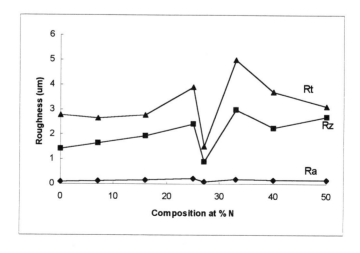

Figure 6 *Roughness of films' dependence on composition*

determined by the target power. Therefore the nitrogen partial pressure control was the main process parameter which influenced the phase composition of the films. As can be seen from Figure 1 the composition of nitrogen was sensitively affected by the nitrogen partial pressure before the nitrogen concentration reached 33 at % N. During this stage, the target was not poisoned so that the sputtering rate of titanium was constant and high plasma density was produced in front of the target. As the consumption of the nitrogen in the plasma was proportional to the nitrogen gas flow during this stage, there was no distinct change in total chamber pressure. It was, therefore, necessary to control the nitrogen partial pressure precisely during the process in order to produce pure Ti_2N coatings. However, the nitrogen partial pressure control was a complicated problem which was influenced by several parameters such as target sputtering rate, loading of the chamber, substrate bias voltage, nitrogen gas flow and consumption of nitrogen during the process. These inter-related parameters affected the reproducibility of pure Ti_2N films. From the work presented in this paper, the nitrogen partial pressure must be precisely controlled in the range of 2.0×10^{-5} mbar in order to produce pure Ti_2N coating. If the nitrogen partial pressure is below or above that range, the $\alpha Ti(N)$ or δTiN will be present in the coating respectively, and the Ti_2N phase disappears.

The pure Ti_2N coating was not as hard as stoichiometric TiN coatings. However the adhesion as measured using Rockwell test was excellent. Ti_2N coatings up to $15\mu m$ still have excellent adhesion ($R_c = 1$). In addition the Ti_2N coatings achieved a relatively high hardness ($22000N/mm^2$) and very low roughness which are useful for wear resistant applications

5 CONCLUSION

1. The development of the coating phase structure was critically dependent on the concentration of nitrogen within the plasma
2. The optimised concentration of nitrogen for Ti_2N film was 27 at %N. Any deviation from this composition resulted in either the $\alpha Ti(N)$ phase or the stoichiometric TiN phase coexisting with the Ti_2N phase.
3. Highest hardness was achieved in film containing approximately 40 at %N. The hardness of Ti_2N film was relatively lower than that of the stoichiometric TiN film. However the adhesion of the Ti_2N coating was excellent.
4. The pure Ti_2N film was exceptionally smooth. The low roughness and relatively high hardness might make it an ideal candidate for machine tool applications.

Acknowledgement

S. Yang acknowledges a MRI research student bursary.

References

1. J. E. Sundgren, *Thin Solid Films*, 1985, **128**, 21.
2. M. Kiuchi, K. Fujii, H. Miyamura, K. Kadono, M. Satou, and F. Fujimoto, *Nuclear Instruments and Methods in Physics Research*, 1989, **B37/38**, 701.

3. M. Hansen, 'Constitution of Binary Alloys', McGraw Book Co., Inc. New York, 1st edi., 1958.

4. V. Poulek, J. Musil, V. Valvoda and L. Dobiasova, *Materials Science and Engineering*, 1991, **A140**, 660.

5. M. Van Stappen, K. Debruyn, C. Quaeyhaegens, L. Stals, and V. Poulek, *Surf. Coat. Technol.*, 1995, **74/475**, 143.

6. William D. Sproul, *Surf. & Coat. Technol.*, 1987, **33**, 73.

7. A.K. Vershina, S.D. Izotova, *Fizika i Khimiya Obrabotki Materialov*, **N3**, May-Jun. 1991, 65.

8. W. D. Sproul and P. J. Rudnik, *Thin Solid Films*, 1989, **171**, 171.

9. W. D. Sproul, P. J. Rudnik and M. E. Graham, *Surf. Coat. Technol.*, 1989, **39/40**, 355.

10. S. Kadlec, J. Musil, W. Münz and V. Valvoda, Pro. 7th Int. Conf. on Ion and Plasma Assisted Technol., Genf, Switzerland, CEP Consultants, Edinburgh, 1989, p. 100.

11. L. Hultman, W. Münz, J. Musil, S. Kadlec, I. Petrov, J. E. Greene, *J. Vac. Sci. Technol.*, 1991, **A9**, 434.

12. A. Matthews and D. G. Teer, *Thin Solid Films*, 1980, **72**, 541.

13. B. E. Jacobson, R. Nimmagadda and R. F. Bunshah, *Thin Solid Films*, **63**, 333.

14. W. Münz, D. Schulze, F. J. M. Hauzer, *Surf. Coat. Technol.*, 1992, **50**, 169.

15. V. Hoffmann, Fresenius *J. Anal Chem.*, 1993, **346**, 165.

16. M. Ives, J. Cawley and J. S. Brooks, *Surf. Coat. Technol.*, 1993, **61**, 127.

17. 'Powder Diffraction File', JCPDS International Centre for Diffraction Data, Swarthmore, PA 1991.

18. V. Poulek, J. Musil, V. Valvoda, and R. Cerny, *Journal of Physics*, 1988, **D21**, 1657.

2.1.3
Evaluation of Defects in TiN Thin Films by Various Electrochemical Techniques

Seiji Nitta[1] and Yuji Kimura[2]

[1]GRADUATE SCHOOL, KOGAKUIN UNIVERSITY, 1–24–2 NISHISHINJUKU, TOKYO 163–91, JAPAN

[2]DEPARTMENT OF CHEMICAL ENGINEERING, FACULTY OF ENGINEERING, KOGAKUIN UNIVERSITY, 1–24–2 NISHISHINJUKU, SHINJUKU-KU, TOKYO 163–91, JAPAN

1 INTRODUCTION

Much dry coating technologies by CVD and PVD method have been studied in these days for the purpose of improving corrosion resistance of stainless steel[1]. For example, TiN (Titanium Nitride), which has high hardness, abrasion-resistance and colour as gold, is worthy of consideration as a coated material for corrosion resistance in addition to some superior functions on metal substrate surface. However, corrosion resistance of TiN-coated stainless steel by dry coating process is strongly influenced by the existence of various defects in the thin films, such as pinhole defects through which corrosive media reach the substrate directly[2,3]. Therefore, the present authors have been investigating the pinhole-ratio in ceramic coated films by conducting various tests employing electrochemical methods[4,5]. For evaluating pinhole-ratio by electrochemical methods, it is necessary, to confirm the correspondence between the defect region in coated film and the penetrated area to cause substrate corrosion in electrochemical test, because of utilizing exact current density and potential which correlate closely with defect size and number. Therefore in this study, the particular defects were observed in detail before and after the corrosion tests. From the classification of corroded morphologies, the most suitable electrochemical method was chosen to characterize TiN coating film.

2 EXPERIMENTAL PROCEDURES

2.1 Materials

The specimens used for this study were TiN ceramic films made by dynamic ion mixing and plasma CVD. The thickness of the TiN thin films was about 1μm for dynamic ion mixing and about 2μm or 5μm for plasma CVD. The processing conditions of TiN coatings are given in Tables 1 and 2. The chemical compositions of type AIS1304 stainless steel substrate are shown in Table 3. After the lead wire was laid on the specimen, the entire surface, except for 1cm^2 of testing area, was covered with silicon.

2.2 Electrochemical Measurements

The pinhole-ratio has been determined using various kinds of electrochemical methods.

Table 1 *Conditions of TiN coating by dynamic ion mixing*

Base pressure (Pa)	6×10^{-4}
N_2 gas pressure (Pa)	4×10^{-3}
N_2 ion beam energy (keV)	20
N_2 ion beam current (A/m^2)	1.0
N_2 irradiation tline (sec)	500
Ti evaporation rate (m/sec)	1×10^{-9}
Acc. V (keV)	20
Dec. V- (keV)	0.3
Ion beam current (A/m^2)	0.4
YEW current (A)	0.011
Time (sec)	984
Thickness (m)	1×10^{-6}
Substrate temperature (K)	573

That is, these studies were conducted by polarization curve method, critical passivation current density (CPCD) method and electrochemical measurements under coupling condition between specimen and the counter electrode of Pt in which freely corroding condition was almost kept. The setup of the test apparatus is shown in Figures 1 and 2. Electrolytes for each measurement were 0.5kmol/m^3-H$_2$SO$_4$, 0.5kmol/m^3-H$_2$SO$_4$ + 0.05kmol/m^3-KSCN and 0.5 kmol/m^3-H$_2$SO$_4$ + 0.1kmol/m^3-KSCN. Incidentally, for the purpose of causing sufficient anodic dissolution at various defects in coated films, electrolytes containing KSCN aqueous solution were used in the case of CPCD method and measurements were done under coupling condition. Consequently, KSCN concentration was changed slightly by the test. The corroded morphlogies of coating films surface were examined in detail by optical microscope, scanning electron microscope (SEM) and atomic force microscope (AFM).

3 RESULTS AND DISCUSSIONS

3.1 Pinhole-Ratio of TiN Coated Stainless Steels

Figure 3 shows the polarization curves related to the three kinds of TiN films and stainless steel substrate in 0. 5kmol/m^3-H$_2$SO$_4$ aqueous solution. First, evaluation method for defects ratio (pinhole-ratio) was proposed under transpassivity state on the basis of method of pitting potential measurement for stainless steel[6] . That is, by comparing the pitting potentials at which the current density came to be i = 1A/m^2, the pinhole-ratio in coating films could be

Table 2 *Conditions of TiN coating by plasma-CVD*

Pressure (Pa)	133.3
Substrate temperature (K)	973
RF power (W)	400
Gas flow (%)	N_2:79 H_2:19 TiCl$_4$:2

Table 3 *Chemical composition of SUS304 (wt. %)*

C	Si	Mn	P	S	Ni	Cr
0.05	0.43	1.18	0.032	0.025	8.66	18.37

deduced. Judging from two TiN films by plasma CVD, the corrosion resistance of the 5μm thick film was superior to that of 2μm, because the pitting potential of 5μ m was swept in more noble direction compared with that of 2μm. Generally speaking, as the corrosion resistance of ceramic coating films was dependent upon the defect ratio in films, the pitting potential gives rough guide to the evaluation of defects in coating films. Therefore, real pinhole-ratio evaluation could be conducted through comparing defect size with corrosion pit size; we will discuss in next section about this point.

Then, the CPCD method was conducted by measuring polarization curve in 0. 5kmol/m³-H_2SO_4 aqueous solution containing 0.05kmol/m³-KSCN aqueous solution, at which potential was swept from −0.5 to + 0.4V (Ag/AgCl). The values of active peak were more large as shown in Figure 4, compared with the above-mentioned polarization curve method. In the case of CPCD method the pinhole-ratio R was calculated by the ratio of critical passivation current density of TiN-coated stainless steel to that of stainless steel substrate under active state region by equation {1}[2,3], so that the pinhole-ratio in TiN thin films can be determined quantitatively as shown in Table 4.

$$R = F \times \frac{i_{crit}(Coating / Substrate)}{i_{crit}(Substrate)} \times 100\% \qquad \{1\}$$

i_{crit} (Coating/Substrate) : Critical passivation current density of TiN-coated stainless steel
i_{crit} (Substrate) : Critical passivation current density of stainless steel substrate
F: A coefficient of corrosion pit morphology (For hemispherical corrosion pit; F = 1/2)

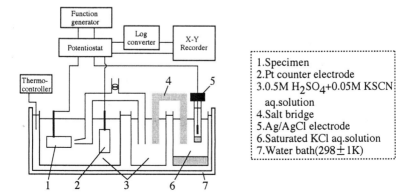

Figure 1 *Schematic block diagram for potential sweeping electrochemical measurements*

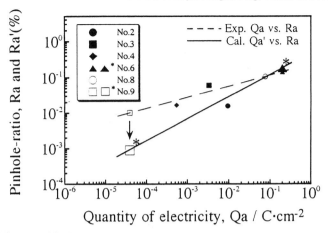

Figure 2 *Schematic block diagram for electrochemical measurements under the coupling condition between specimen and counter electrode of Pt*

The results of potential and current density measurement tests under coupling conditions with counter electrode of Pt are shown in Figure 5. Sudden drops in potential values accompanying current density value increases were recognized within two hours. Times until active state and quantities of electricity depend on the difference of pinhole-ratio in TiN films. In this coupling test method, the pinhole-ratio Ra of coated films can be determined quantitatively from the quantity of electricity Qa through evaluating corrosion pit morphology as shown in Figure 6 (a broken line). Also, in Figure 6, the value of Ra in the smaller region is corrected as Ra' to Qa' calculated by Faraday's law (solid line), because the pinhole-ratio Ra determined by optical microscope observation contains the defects which do not participate in substrate corrosion. The classification of corrosion damage morphologies observed in detail by AFM as shown in Figure 10 should account for the accuracy of pinhole-ratio Ra'.

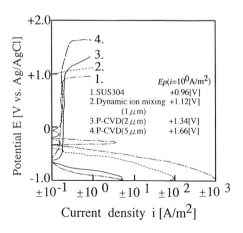

Figure 3 *Polarization curves of specimens in 0.5 kmol/m³-H₂SO₄ aq. solution at 303K, sweep rate 0.33mV/sec*

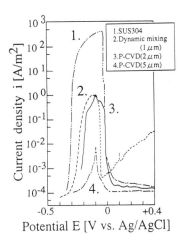

Figure 4 *Anodic polarization curves of specimens in 0.5 kmol/m³-H₂SO₄ + 0.05 kmol/m³-KSCN aq. solution (CPCD method) at 298K, sweep rate 0.33 mV/sec*

3.2 Morphologies of Corrosion Damages in TiN Films

At first, the microstructures of TiN films were examined by optical microscope and AFM before corrosion tests. Morphologies of the defects in the films are classified into two types. Some defects in TiN films, prepared by dynamic ion mixing, are non-penetrating form as shown in Figure 7(a), at which corrosion solution doesn't reach directly to the substrate. Others are pinhole form as shown in Figure 7(c), in which substrate metal was exposed to the solution. On the other hand, the TiN films by plasma CVD have little defects which reach to substrate and have the size over about 10µm. The morphologies of the defects observed by AFM are shown in Figure 8.

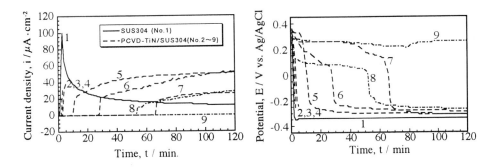

Figure 5 *Change of current density and potential of specimens under coupling conditions in time*

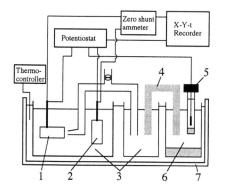
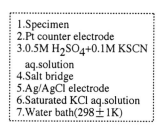

Figure 6 *Relationship between quantity of electricity Qa and pinhole-ratio Ra or Ra'*

The corroded morphologies after electrochemical tests are shown in Figure 7(b) and (d), Similar corrosion damages were generated at the defects with different forms before the tests. That is, this result gives good correspondence between the defect region in coated film and the penetrated area to cause substrate corrosion by electrochemical test. Therefore, the TiN-coated steels prepared by dynamic ion mixing can be examined by using potential sweeping electrochemical measurements, such as the critical passivation current density (CPCD) method and polarization curve method, because there is sufficient adhesive strength between the substrate and the coated films.

However, in the case of TIN-coated stainless steels prepared by plasma CVD, potential sweeping electrochemical measurements bring about a little peeling off in coated film as shown Figure 9(a)(b). Therefore, the TiN-coated stainless steels prepared by plasma CVD can be evaluated by electrochemical measurements under the coupling condition between the specimen and the counter electrode of Pt, because there is little spalling-off of the coated film in this evaluation method. AFM micrographs of corrosion damage observed in the TiN film by plasma CVD after electrochemical measurements under coupling condition are shown in Figure 10. A defect which participates in substrate corrosion was kept at an original form as shown in the Figure 10(a).

Table 4 *Pinhole-ratio of TiN films*

Materials	Pinhole-ratio R(%)
Dynamic ion mixing (1μm)	0.104
Plasma-CVD (2μm)	0.104
Plasma-CVD (5μm)	0.00073

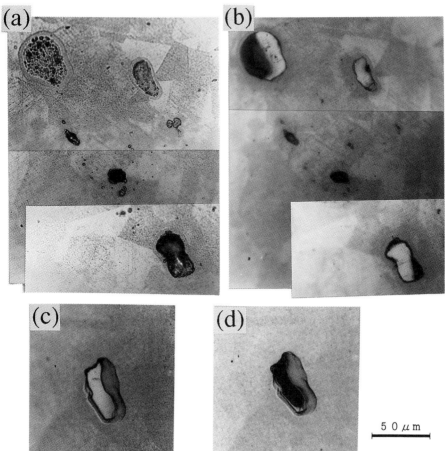

Figure 7 *Morphologies of TiN films by dynamic ion mixing (a) , (c) before electrochemical tests, (b) after polarization curve test (d) after CPCD test*

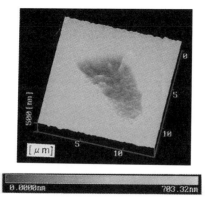

Figure 8 *Morphologies of TiN films by plasma-CVD before test*

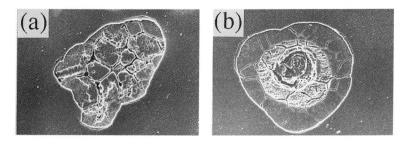

$20 \, \mu m$

Figure 9 *Morphologies of TiN films of 5µm thickness bv plasma-CVD (a) after polarization curve test (b) after CPCD test*

4 CONCLUSION

Evaluation of the pinhole-ratio in several TiN thin films was conducted by using various electrochemical methods. The appearance of defects in TiN thin films was examined by optical microscope, SEM and AFM.

The results obtained are summarized as follows :

1. From corroded morphologies to defects, the applicable electrochemical method for TiN coated film should be chosen.
2. TiN-coated stainless steels prepared by plasma CVD are best evaluation by electrochemical measurements under the coupling condition between the specimen and the counter electrode of Pt, because there is little spalling-off of the coated film in this evaluation method.
3. TiN-coated stainless steels prepared by dynamic ion mixing can be examined by using potential sweeping electrochemical measurements, because there is sufficient adhesive strength between the substrate and the coated films.

Figure 10 *Morphologies of corrosion damage in pinhole-ratio low film observed after electrochemical measurements under coupling condition, (a) pinhole-defect, (b) non-penetrating defect and (c) swelling defect*

References

1. H. Takeda, *Ceramic coating*, Nikkankogyo,Tokyo, 1988.
2. K. Sugimoto, 95th corrosion protenction symposium, 1993, 1.
3 . S. Nitta and Y. Kimura, *The Japan Society of Mechanical Engineering*, 1995, **61**, 1914.
4. T. Osaka and N. Koyama, 'Electrochemical Method', Koudansya, Tokyo, 1990, Chapter 5.
5. JSNE standard S-010, 1996.
6. JIS G 0577. 'Method of Pitting Potential Measurement for Stainless steel', 1981.

2.1.4

Annealing of Defects in PVD TiN: a Positron Annihilation Spectroscopy Study

S. J. Bull[1], A. M. Jones[2], A. R. McCabe[2], A. Saleh[3] and P. Rice-Evans[3]

[1]MATERIALS DIVISION, NEWCASTLE UPON TYNE, UK.

[2]AEA TECHNOLOGY, B552 HARWELL LABORATORY, OXFORDSHIRE, UK.

[3]DEPT OF PHYSICS, ROYAL HOLLOWAY, UNIVERSITY OF LONDON, EGHAM, SURREY, UK.

1 INTRODUCTION

It is well recognised that the inter-relationship between the properties of thin films and coating microstructure, which may be manipulated to a certain extent by process parameters, is of crucial importance if coatings are to be optimised for a range of applications[1-4]. Many of the properties of physical vapour deposited (PVD) and chemical vapour deposited (CVD) coatings change with temperature so it is essential to know what structural changes are likely at the operating temperature if the best coatings are to be developed.

For PVD coatings tempering at elevated temperatures (up to 900°C) results in a reduction in the level of internal stress and hardness[5-8]. Similar reductions in stress, lattice parameter and resistivity have been reported for sputtered coatings deposited at low temperature and annealed at temperatures as low as 300°C[9,10].

At least some of these differences arise from microstructural changes which occur on heating[8]. For PVD TiN on steel the coating at room temperature exhibits a compressive residual stress which is a combination of ion bombardment – induced growth stress and thermal stress due to the expansion mismatch between coating and substrate. On heating a tensile thermal stress is induced which counteracts these compressive stresses and as the temperature is increased the overall stress level in the coating is reduced. At the deposition temperature only the growth stresses remain – further heating allows these also to be counteracted until the stress falls to zero when the columns which comprise the microstructure are pulled apart. This is partly due to creep in the steel substrate. However, dramatic changes in lattice parameter and microstrain broadening occur at temperatures above the deposition temperature which cannot simply be rationalised in terms of gross microstructural changes. Rather these changes represent the annealing of individual defects within the columns which comprise the coating.

There is some discussion about the nature of defects introduced in PVD films by ion bombardment and how these lead to generation of residual stress[11,12]. In this study we have used positron annihilation spectroscopy to ascertain the effect of annealing on the vacancies in the films and hence determine what point defects contribute to stress generation. The results are compared to those from high temperature X-ray diffraction studies on identical samples.

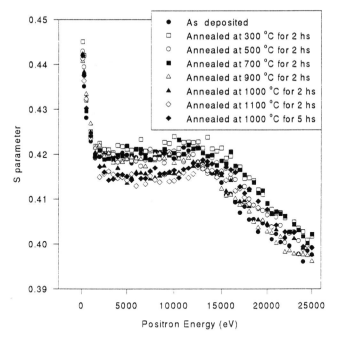

Figure 1 *Variation of S-parameter with positron energy for unimplanted TiN as a function of heat treatment*

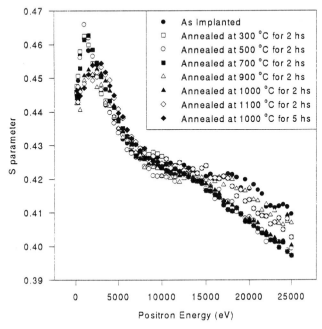

Figure 2 *Variation of S-parameter with positron energy for argon implanted TiN as a function of heat treatment*

2 EXPERIMENTAL

Titanium nitride coatings were deposited onto stainless steel coupons by sputter ion plating[13]. The coatings were deposited with a bias voltage of −90V to a nominal thickness of 2–3 μm, at a deposition temperature of 500°C. To enhance the number of defects in the coating some samples were ion implanted with Argon or nitrogen to a dose of 1×10^{17} ions/cm^2 at an energy of 60keV. TRIM calculations[14] indicate that approximately equal numbers of vacancies and interstitials are created under these conditions.

Both implanted and unimplanted samples were annealed in a vacuum furnace with a base pressure of better than 1×10^{-6} Torr so little or no oxidation occurred. Samples were heated and cooled under vacuum and spent 1h at the annealing temperature which varied from 300°C to 1100°C.

The samples were then analysed in the variable energy TACITUS positron beam at Royal Holloway. In this apparatus positrons from ^{22}Na are moderated by a tungsten mesh, guided magnetically towards the sample chamber and then focused by an electrostatic lens onto the target. By varying the target voltage it is possible to vary the positron energies and hence their mean penetration depth. A germanium detector placed behind the target records the spectrum of the 511keV annihilation photons for 1h at each positron energy. To analyse the spectrum the Doppler-broadening S-parameter was used which is sensitive to the material composition and vacancy concentration. It is defined as the counts in the central region of the annihilation peak divided by the total counts in the peak[14]. In effect positrons may annihilate at the surface,

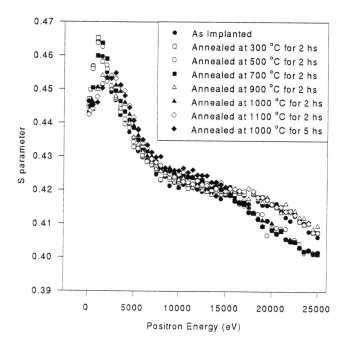

Figure 3 *Variation of S-parameter with positron energy for nitrogen implanted TiN as a function of heat treatment*

in the film or in the substrate and a mathematical model has been successfully employed to account for this in PVD coatings previously[15]. The X-ray diffraction measurements were made using the high temperature attachment on the Harwell-designed APEX goniometer at AEA Technology. The experimental procedure has been described previously[8].

3 RESULTS

Positron annihilation spectra for the unimplanted and annealed samples are shown in Figure 1. The high S-parameter at low positron energy represents annihilation at the surface whereas the following plateau is due to annihilation in the coating and the slope at the highest positron energies is due to the effect of the substrate. It is clear that the S-parameter remains approximately constant for annealing up to 900°C, but is reduced above this temperature.

For the ion implanted samples the vacancies produced by ion implantation generate a peak close to the surface of the TiN layer at low positron energies (Figures 2 and 3). There is little change in the height or position of this peak for annealing below 900°C, but at this temperature the peak height is reduced and its maximum moves to higher positron energies. This is evidence that some vacancy annealing has occurred, probably by diffusion of vacancies to the free surface. The change at 900°C is rather dramatic, but somewhat surprisingly the annealing appears to be less marked at higher temperatures which is not expected for a thermally activated process.

The high temperature X-ray diffraction studies on unimplanted coatings show that the residual stress in the coating is reduced to zero at 800°C (Figure 4). At this temperature and above considerable annealing of lattice strain occurs (Figure 5) as the coating columns are pulled apart[8] dramatically increasing the area of free surface available to act as a sink for the defects to disappear at. It thus seems likely that at 900°C the vacancies migrate to this new

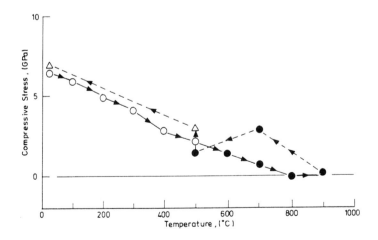

Figure 4 *Variation of compressive residual stress with annealing temperature. The complete thermal cycle consists of room temperature to 500°C hold for 15h (o), 500°C to 900°C and back to 500°C hold for 15h (•), and finally 500°C back to room temperature (Δ)*

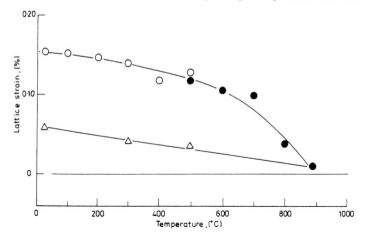

Figure 5 *Variation of lattice strain for the (422) peak of TiN as a function of annealing temperature. Symbols as in Figure 4*

surface where they disappear causing the rapid change in S-parameter observed at this temperature. Vacancies can also diffuse to other traps (defect clusters, dislocations etc.) where they do not disappear and it is only those that are close enough to the surface which are lost at this point. This would also explain the movement in the implantation peak maximum.

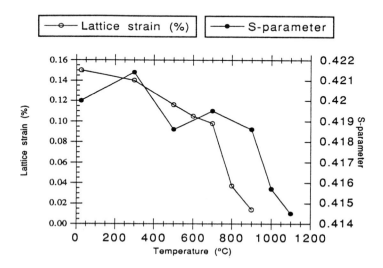

Figure 6 *Comparison between S-parameter and lattice strain as a function of annealing temperature*

4 DISCUSSION

Careful comparison of the change in S-parameter and microstrain broadening with annealing temperature shows that the microstrain broadening decreases continuously with temperature and begins to change dramatically at 800°C whereas the S-parameter does not change appreciably until 900°C when it drops rapidly (Figure 6). This would imply that the changes in the X-ray diffraction measurements are not caused by vacancies which can only contribute at the highest temperatures. It is well known that interstitials are mobile at lower temperatures than vacancies[16] and it has been suggested that the change in stress and lattice parameter in the annealing of low temperature deposited TiN is due to the movement of interstitials[10]. The work reported here would tend to confirm this suggestion and indicates that interstitials will also be important for films deposited at high temperature. Indeed taking into consideration the relative lattice distortions caused by an interstitial atom or a vacancy (Figure 7), it can easily be seen that X-ray diffraction will be more sensitive to the distortions created by an interstitial since these extend over a much greater volume than those for a vacancy. In both cases there is a tendency to increase the lattice parameter which is what is generally observed by X-ray diffraction, though the change associated with the vacancy is too small to be significant except at very high vacancy concentrations.

The origin of the residual stress produced by ion bombardment has been the subject of some discussion in the literature, though it is known to be related to the energy or momentum transfer from the bombarding ions[17]. However, from the work reported here it would seem likely that it is the creation of interstitials leading to lattice expansion which is responsible for the observed stresses. Constraint from the neighbouring columns in the coating is necessary to generate the stress otherwise this can be relaxed by expansion of the column into the space around it. Since ion bombardment is usually sufficiently energetic to create subsurface defects the interstitials become lodged into the structure until sufficient reduction of this constraint occurs, usually by heating to greater than the deposition temperature. It is for this reason that significant stress or lattice parameter relaxation is only observed for annealing at temperatures

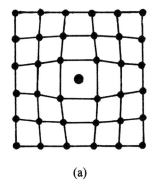
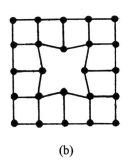

(a)   (b)

Figure 7 *Schematic of the lattice distortions due to (a) an interstitial and (b) a vacancy*

Table 1 *Relative sizes of interstitials and the sites which they could occupy*

Interstitial	Radius r_i (nm)	Site Radius r_s (nm)	r_i/r_s
Argon interstitial on octahedral titanium	0.154*	0.15	1.03
Titanium interstitial on octahedral nitrogen site	0.15	0.0621	2.4
Nitrogen interstitial on tetrahedral site	0.0621	0.0315	2.0

* Ionic radius

above the deposition temperature[8,10,18,19].

There are three types of interstitial which could contribute to the observed behaviours:

1.   Implanted argon on octahedral titanium sites.
2.   Titanium (or argon) interstitials on octahedral nitrogen sites.
3.   Nitrogen interstitials on tetrahedral sites.

Under the deposition conditions used here most of the argon will come to rest in the outermost layer of the coating from where it can easily escape. There is no measurable trapped argon in the as-deposited films and certainly not enough to explain the increase in lattice parameter associated with the growth stresses. TRIM calculations indicate that at 100eV some 1.6 vacancies/ion are created (and presumably the same number of interstitials) with about 50% more nitrogen vacancies than titanium vacancies. It is probable that both types of interstitial are responsible for the observed lattice expansion behaviour.

An indication of the effect of each of the interstitials on the lattice strain in the host TiN lattice can be gained from consideration of the size of the interstitial and the size of its potential location in an undeformed lattice. Table 1 summarises these values for the cases introduced above. The argon interstitial has only a small effect on the titanium site. It would appear that the titanium (or argon) interstitial on a nitrogen site will generate more strain than the nitrogen interstitial on the tetrahedral site by this approach. However, given the increased numbers of nitrogen vacancies (and hence interstitials) created during coating both defects are likely to be important. What argon that does become entrapped in the coating will also have an additional effect.

One feature which is commonly reported for PVD TiN films is that the unit cell dimensions of the nitride determined for each (*hkl*) reflection can vary, and that these deviations are a sensitive function of internal stress and thickness[20] and deposition rate[21]. These pseudomacrostrains are completely relaxed by heating to 900°C – the change in lattice spacing of the {111} planes occurs rapidly above 600°C[8] in a similar temperature range to the change in lattice strain reported here. These pseudomacrostrains have been attributed to argon incorporation[22] and nitrogen occupying tetrahedral sites in the TiN lattice[23]. The results presented here would support both these hypotheses, but the relaxation process is critically dependent on the interaction of the columnar boundaries as suggested previously[13]. To completely remove an interstitial from the system it must either recombine with a vacancy, or diffuse to a free surface. Given that the vacancies and interstitials produced by ion bombardment can become quite widely separated (e.g. by focused collision sequences) and vacancies may be lost from the system in other ways (e.g. by condensation into dislocation loops), the diffusion of interstitials to a free surface is the most attractive method for their removal. Further work is underway to assess the pseudomacrostrains in the ion implanted TiN coatings produced here in an attempt to clarify the role of argon and nitrogen interstitials.

5 CONCLUSIONS

Positron annihilation spectroscopy is a powerful technique for studying vacancies in bulk materials and coatings. The results obtained in this study have highlighted the importance of interstitials in the generation of lattice strain and residual stress in PVD TiN coatings. Relaxation of lattice strain can occur at low temperatures but is most significant at temperatures above the deposition temperature when tensile thermal stresses and creep in the substrate start to open up the coating microstructure. A practical consequence is that PVD coatings should ideally be used at service temperatures below the deposition temperature if major relaxation or changes in properties are to be avoided.

To minimise the lattice strain and residual growth stress on PVD films it is thus necessary to reduce the number of interstitials produced during coating. This can be achieved by reducing the mass of the sputtering gas (with penalty on coating rate), increasing the mass of coating atoms (a restrictive approach) or reducing the energy and increasing the flux of ion bombardment during deposition. This latter approach provides the most practical solution to minimising stress and underlies the drive to increase the ion current density at the substrates in most PVD coating processes.

Acknowledgements

The work report here was supported by the EPSRC and the Corporate Research Programme of AEA Technology.

References

1. J. E. Sundgren, B O Johansson and S E Karlsson, *Thin Solid Films*, 1983, **105**, 353.
2. D. S. Rickerby and P. J. Burnett, *Thin Solid Films*, 1988, **157**, 195.
3. D. S. Rickerby and S. J. Bull, *Surf. Coat. Technol.*, 1989, **39/40**, 315.
4. S. J. Bull, *Vacuum*, 1992, **43**, 679.
5. D. T. Quinto, G. J. Wolfe and P. C. Jindal, *Thin Solid Films*, 1987, **153**, 19.
6. A. J. Perry and L. Chollet, *Surf. Coat. Technol.*, 1988, **34**, 123.
7. A. J. Perry, *Thin Solid Films*, 1987, **146**, 165.
8. D. S. Rickerby, S. J. Bull, A. M. Jones, F. L. Cullen and B. A. Bellamy, *Surf. Coat. Technol.*, 1989, **39/40**, 397.
9. C. Ernsberger, A. J. Perry, L. P. Lehman, A. E. Miller, A. R. Pelton and B. W. Dabrowski, *Surf. Coat. Technol.*, 1988, **36**, 605.

10.   F. Elstner, H. Kupfer and F. Richter, *Phys. Stat. Sol. (a)*, 1995, **147**, 373.
11.   I. Goldfarb, J. Pelleg, L. Zevin and N. Croitoru, *Thin Solid Films*, 1991, **200**, 117.
12.   O. Knotek, R. Elsing, G. Kramer and F. Jungblat, *Surf. Coat. Technol.*, 1991, **46**, 265.
13.   D. S. Rickerby and R. B Newbery, *Vacuum*, 1988, **38**, 161.
14.   P. J. Schultz and K. G. Lynn, *Rev Mod. Phys.*, 1988, **60**, 701.
15.   S. J. Bull, P. C. Rice Evans and A. S. Saleh, *Surf. Coat. Technol.*, 1996, **78**, 42.
16.   M W Thompson, 'Defects and Radiation Damage in Metals', Cambridge University Press, 1969.
17.   S. J. Bull, A. M. Jones and A. R. McCabe, *Surf. Coat. Technol.*, 1992, **54,55**, 173.
18.   A. J. Perry, *J. Vac. Sci. Technol.*, 1988, **A6**, 2140.
19.   F. Elstner, A. Ehrlichs, H. Giegengack, H. Kupfer and F. Richter, *J. Vac. Sci. Technol.*, 1994, **A12**, 476.
20.   D. S. Rickerby, A. M. Jones and B. A. Bellamy, *Surf. Coat. Technol.*, 1989, **37**, 111.
21.   A. J. Perry and J. Schoenes, *Vacuum*, 1986, **36**, 149.

Section 2.2 Thermal, Plasma, Weld and Detonation

2.2.1
Spouted Bed Reactor for Coating Processes : a Mathematical Model

G. Mazza[1], I. Sanchez[2], G. Flamant[2] and D. Gauthier[2]

[1]ON LEAVE OF ABSENCE FROM THE UNIVERSITY OF LA PLATA AND CONICET (ARGENTINA)

[2]INSTITUT DE SCIENCE ET DE GÉNIE DES MATÉRIAUX ET PROCÉDÉS, IMP-CNRS, BP 5 ODEILLO, F–66125 FONT ROMEU CEDEX FRANCE

1 INTRODUCTION

The specific advantages of the fluidised and spouted beds gas-solid processes (great exchange area, homogeneous bed temperature, etc.) favour their use for various coating applications such as growth of bed particles or surface modification of particles or objects immersed in the bed. In this paper we describe a new non-isothermal model developed for a spouted bed reactor. The model is tested on CVD deposition of silicon from silane.

2 MODEL DESCRIPTION

A stable spouted bed can be described as a combination of three domains: a dilute central core (spout) with upward moving solid entrained by a co-current flow of fluid, a dense phase annular region (annulus) with counter-current percolation of fluid, and a fountain where the solid from the spout returns to the annulus.

Only a few spouted bed reactor models can be found in the literature. Mathur and Lim[1] defined a one-dimensional model of a spouted bed reactor (SBR). This isothermal model divides the reactor in two uniform regions called the "spout" and the "annulus". Plug flow is assumed in each region. Piccinini et al[2] and Littman et al[3] tested this model with first order catalytic isothermal reactions. Smith et al[3] presented a non-isothermal version of this SBR modelling based on a pseudo-homogeneous approach.

Piccinini et al[2] proposed an isothermal model based on the hydrodynamic stream-tube approach developed by Lim and Mathur[4]. This approach gives a better description of particle and gas flows in the annulus than the one-dimensional model, and it is thus expected to be better for the design of large reactors.

In 1992, Hook et al[5] presented a non-isothermal bi-dimensional model which included for the first time the fountain region. The flow of gas and solid was calculated by solving momentum balances.

We developed a new non-isothermal SBR model based on Piccinini et al's[2] isothermal model. The stream-tube approach allows easy determination of gas flow thanks to known correlations and it gives a good description of gas and particle circulation[4,6–8]. Heat transfer is included by means of an heterogeneous approach. Mass transfer and chemical reaction effects are also taken into account. The model divides the bed in three uniform regions according to their voidage and hydrodynamic properties. These regions are the spout, the annulus and the fountain. The model is illustrated in Figure 1.

2.1 Definition of the Regions

2.1.1 Spout. The model can stand a variable radius of the spout (rs) with the vertical coordinate z. We choose rs constant (mean value) according to the correlation quoted by Hook et al[5]:

$$\frac{\langle D_s \rangle}{D_c} = \left(\frac{2.08 \rho_g U_i'^2}{(\rho_p - \rho_g)(1 - \varepsilon_{mf})g D_c} \right)^{\frac{1}{4}}$$

2.1.2 Fountain. The fountain is divided in two regions of high porosity: the fountain core and the down-flowing fountain. The down-flowing fountain receives poor gas from the annulus where most of the reaction occurs. On the contrary, the fountain core receives the gas still rich in reactants from the spout (low conversion is reached in the spout). Therefore, the down-flowing fountain region can be neglected in comparison to the fountain core. The shape of the fountain is determined by Hook et al's[5] procedure. The fountain core diameter is supposed to increase linearly with distance; the fluid leaving the spout at $z = H_c$ remains within the fountain core. The fluid superficial velocity at the top of the fountain is assumed to be equal to

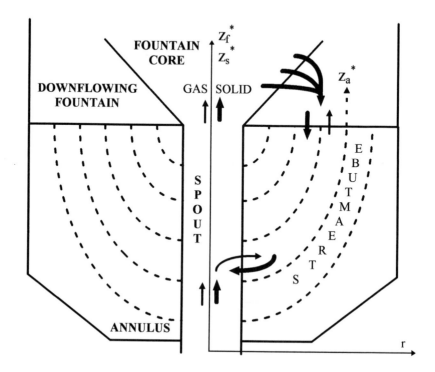

Figure 1 *Schematic representation of a spouted bed reactor*

the inlet superficial velocity, U'_{in} when $U'_{in} \geq U_{mf}$, or to the minimum fluidizing velocity U_{mf} when $U'_{in} < U_{mf}$. These hypotheses permit the fountain core extend determination.

$$U'_{in} = \frac{\pi D_{in}^2 U_{in}}{\pi D_c^2} \text{ and } r_f(H_f) = \sqrt{\frac{W_{fg}}{\pi U_f(H_f)\rho_g}}$$

$$r_f(z) = r_s(H_c) + \left(\frac{r_f(H_t) - r_s(H_c)}{H_f}\right)(z - H_c)$$

2.1.3 Annulus. The stream-tube approach defined by Lim and Mathur[4] is used in the annulus. This zone is treated as a set of stream-tubes, and the stream-tube boundaries are determined according Lim and Mathur's[4] method as modified by Piccinini[2].

2.2 Hydrodynamic Hypotheses

The pressure is supposed to be constant all over the reactor. Plug flows with cross flows of solid and gas are considered in the spout and in the fountain. Every annulus streamtube is a plug flow reactor with no dispersion; mixing does not occur radially between streamtubes or axially along a stream-tube.

2.2.1 Velocities Determination. All velocities and flow-rates for both particle and gas phases are estimated from existing correlations and mass balances. Consequently, momentum equations are not included in the model formulation.

A modified version of Mamuro and Hattori's equation[9] proposed by Piccinini et al[2] is used to account for the gas distribution between the spout and the annulus[2,9].

$$W_{ag(total)} = 0.88 U_{mf} \frac{\pi}{4}\left(D_c^2 - D_s^2\right)\left[1 - \left(1 - \frac{z}{H_m}\right)^3\right]\rho_g$$

Since the particle circulation rate cannot be predicted with certainty, as a first approximation, the interstitial particle velocity in the spout (V_s) is assumed to be proportional to a slip fraction γ, to the interstitial gas velocity (U_s) on the top of the spout[3]. The particle cross flow velocity (V_r) is determined from solid continuity equation.

$$V_s(z) = \gamma \frac{(1 - \varepsilon_s(z))}{\varepsilon_s(H_c)} U_s(H_c) \quad , \quad \gamma = 0.3 \text{ and } U_s = \frac{\left(W_{in} - W_{ag(total)}\right)}{\pi r_s^2 \rho_g}$$

To evaluate the particle flow-rate in the fountain, the relation proposed by Hook et al[5] is used:

$$\frac{W_{fp}(z)}{W_{sp}(z)} = \left[1 - \left[\frac{z - H_c}{H_t - H_c}\right]^2\right]^{\frac{1}{2}} \text{ and } V_f(z) = \frac{W_{fp}(z)}{\pi r_f^2(z)\rho_p}$$

In the annulus region, the gas and particle flow-rates corresponding to each stream-tube are deduced from mass balances in the spout. Then the velocities are calculated knowing the conical flowing cross-section area at any position along the streamtube.

2.2.2 Porosity. We took the expression proposed by Smith et al (1982) for the porosity variation in the spout[3], based on Littman et al[10] and Lefroy et al[11] results.

$$\varepsilon_s(z) = 1 - \beta \frac{z}{H_c} \quad \text{with } \beta = 0.2$$

For the fountain porosity, the approach developed by Hook et al[5] is adopted, combining both following equations:

$$\frac{[1-\varepsilon(z)]}{[1-\varepsilon(H_t)]} = \left[1 - \left(1 - \frac{z}{H_t}\right)^2\right]^{\frac{1}{2}} \quad \text{and} \quad \frac{[1-\varepsilon(H_c)]}{[1-\varepsilon(H_t)]} = \left[1 - \left(1 - \frac{H_c}{H_t}\right)^2\right]^{\frac{1}{2}}$$

The voidage in the annulus varies in a defined region adjacent to the spout. The length of this region has been chosen equal to about three particle diameters (classical assumption). Out of this region, the porosity is constant and equal to the minimum fluidisation voidage $\varepsilon_{mf} = 0.4$.

$$\begin{cases} r_s \leq r \leq (r_s + 3\,d_p)\,, \quad \varepsilon_a = \varepsilon_s \left(\dfrac{\varepsilon_s}{\varepsilon_a}\right)^{\frac{(r_s - r)}{3 d_p}} \\[3mm] (r_s + 3\,d_p) \leq r \leq r_c\,, \quad \varepsilon_a = 0.4 \end{cases}$$

2.3 Heat Transfer Hypotheses

The reactor can exchange heat with the external medium through its walls, (or it can be adiabatic, $Q_e = 0$).

An heterogeneous approach (gas and solid temperatures are different) is adopted. The fluid particle heat transfer in the annulus is modelled using the packed bed heat transfer correlation of Handley and Heggs (1968)[12], recommended for Re < 500. For the spout and the fountain regions, Rowe and Claxton's (1965) correlation is used[13]. This last correlation was developed especially for high voidage uniform fluid-particle suspensions. For both correlations, packed bed and high voidage, the relative superficial velocities between the fluid and the particles must be used for the spouted bed, because of the particles' motion.

$$\text{Spout and Fountain}: \quad \frac{h_{pg} d_p}{\lambda_g} = \frac{2}{\left[1 - (1-\varepsilon)^{\frac{1}{3}}\right]} + \frac{0.61}{\varepsilon} \, Pr^{\frac{1}{3}} Re^{0.55}$$

$$\text{Annulus}: \quad h_{pg} = 0.255 \frac{\lambda_g}{\varepsilon_a d_p} Pr^{\frac{1}{3}} Re^{\frac{2}{3}}$$

The corresponding conductive heat transfer between adjacent stream-lines is also considered.

The conductive heat transfer coefficient between two adjacent streamlines has been evaluated following the approach given by Mazza and Barreto[14]. λ_{ep} is the effective thermal conductivity estimated from Kunii and Smith's[15] correlation, and Δ is an effective length between the two adjacent stream-lines which correspond to a rhombohedral array configuration. Δ is calculated from Mazza and Barreto[16].

$$h_{lk} = \frac{\lambda_{ep}}{\Delta} \quad \text{and} \quad \Delta = 0.73869 \frac{d_p}{\left(1 - \varepsilon_a\right)^{\frac{1}{3}}}$$

The energy generated by chemical reaction is assigned to gas for gas-phase reactions and to particles for solid surface reactions. The energy is then exchanged between both phases by gas-particle heat transfer.

2.4 Chemical Reaction

Fluid-particle mass transfer resistances are negligible, or are lumped with the kinetic rate expression. The reaction rate is expressed in terms of the bulk gas concentrations. The particles' physical properties are assumed constant : all variations due to the coating and attrition are neglected. The model can be used with any type of reaction in both solid and gas phases, assuming the reaction kinetics framework is known. So far, the total mole number variation in the gaseous phase has not been taken into account in the model formulation.

2.5 Sets of Equations to Solve

The calculations are carried out versus dimensionless coordinates. The spout is described by the following mass and heat conservation equations:

$$
\begin{cases}
U_s H_c \rho_g \dfrac{d\chi_{si}}{dz_s^*} - r_i = 0 \quad , i = 1,.., nreac \\[2ex]
\left(\rho C_p\right)_g U_s H_c \dfrac{dT_{sg}}{dz_s^*} = h_{pg} \dfrac{6\left(1 - \varepsilon_s\right)}{d_p}\left[T_{sp} - T_{sg}\right] + \sum_{i=gas\ reaction}(-\Delta H_{Ri})r_i \\[2ex]
\left(\rho C_p\right)_p \left[V_s H_c \dfrac{dT_{sp}}{dz_s^*} - V_r \dfrac{\pi D_s}{A_s}\left(T_{apk}(z_a^* = 0) - T_{sp}\right)\right] = h_{pg} \dfrac{6\left(1 - \varepsilon_s\right)}{d_p}\left[T_{sg} - T_{sp}\right] + \sum_{i=solid\ reaction}(-\Delta H_{Ri})r_i
\end{cases}
$$

The fountain core can be considered as an extension of the spout; their modelling equations are very similar:

$$
\begin{cases}
U_f H_f \rho_g \dfrac{d\chi_{fi}}{dz_f^*} - r_i = 0 \quad , i = 1,.., nreac \\[2ex]
\left(\rho C_p\right)_g U_f H_f \dfrac{dT_{fg}}{dz_f^*} = h_{pg} \dfrac{6\left(1 - \varepsilon_f\right)}{d_p}\left[T_{fp} - T_{fg}\right] + \sum_{i=gas\ reaction}(-\Delta H_{Ri})r_i \\[2ex]
\left(\rho C_p\right)_p V_f H_f \dfrac{dT_{fp}}{dz_f^*} = h_{pg} \dfrac{6\left(1 - \varepsilon_f\right)}{d_p}\left[T_{fg} - T_{fp}\right] + \sum_{i=solid\ reaction}(-\Delta H_{Ri})r_i
\end{cases}
$$

In the annulus region, conductive heat transfer occurring between two adjacent stream-tubes links all these streamtubes. The annulus equations are :

$$
\left.
\begin{bmatrix}
\left.
\begin{array}{l}
U_{ak}I_k\rho_g\dfrac{d(\chi_{aik})}{dz_a^*} - r_i = 0 \quad, \; i = 1,..,nreac \\[2ex]
\left(\rho C_p\right)_g U_{ak}L_k \dfrac{dT_{agk}}{dz_a^*} = h_{pg}\dfrac{6(1-\varepsilon_a)}{d_p}\left[T_{apk}-T_{agk}\right] + \sum\limits_{i=gas\,reaction}(-\Delta H_{Ri})r_i
\end{array}
\right\} k = 1,..,nmax
\\[4ex]
\left.
\begin{array}{l}
\left(\rho C_p\right)_p V_{ak}L_k \dfrac{dT_{apk}}{dz_a^*} = \dfrac{2\pi r_{(k+1)}}{A_{ak}}h_{l(k+1)}\left[T_{apk}-T_{ap(k+1)}\right] + \dfrac{2\pi r_k}{A_{ak}}h_{lk}\left[T_{apk}-T_{ap(k-1)}\right] \\[2ex]
+h_{pg}\dfrac{6(1-\varepsilon_a)}{d_p}\left[T_{apk}-T_{agk}\right] - \sum\limits_{i=solid\,reaction}(-\Delta H_{Ri})r_i
\end{array}
\right\} k = 2,..,nmax\text{-}1 \\[4ex]
k = 1 \\[1ex]
\left(\rho C_p\right)_p V_{ak}L_k \dfrac{dT_{apk}}{dz_a^*} = \dfrac{2\pi r_{(k+1)}}{A_{ak}}h_{l(k+1)}\left[T_{apk}-T_{ap(k+1)}\right] + h_{pg}\dfrac{6(1-\varepsilon_a)}{d_p}\left[T_{apk}-T_{agk}\right] - \sum\limits_{i=gas\,reaction}(-\Delta H_{Ri})r_i + Q_e \\[2ex]
k = nmax \\[1ex]
\left(\rho C_p\right)_p V_{ak}L_k \dfrac{dT_{apk}}{dz_a^*} = \dfrac{2\pi r_k}{A_{ak}}h_{lk}\left[T_{apk}-T_{ap(k-1)}\right] + h_{pg}\dfrac{6(1-\varepsilon_a)}{d_p}\left[T_{apk}-T_{agk}\right] - \sum\limits_{i=solid\,reaction}(-\Delta H_{Ri})r_i
\end{bmatrix}
$$

2.6 Boundary Conditions

The spout, the fountain and the annulus are represented by different systems of equations and are therefore calculated separately. The boundary conditions link all these three regions. In the spout, the concentration, gas and particle temperatures at the inlet must be known. The temperature profile of the particle leaving the annulus and entering the spout is also needed all over its length to solve the heat transfer equations.

In the annulus, the boundary conditions concern the characteristics of the gas coming from the spout, and the temperature of particles entering on top of the annulus. All the particles flowing out of the fountain are supposed to mix perfectly before falling down on top of the annulus. The resulting uniform temperature $<T_{fp}>$ is calculated by a global heat conservation balance over the fountain volume.

2.7 Numerical Resolution

The non-linear ordinary differential equation systems are solved by using Michelsen's[17] STIFF3 method, as modified by Barreto and Mazza[18] who include the Jacobian matrix numerical evaluation. The STIFF3 package is called three times for solving separately the sets of equations related to the spout, the fountain and the annulus regions, along coordinates

$$z_s^* = \frac{z}{H_c}\;,\; z_f^* = \frac{z}{H_f} \; \text{and} \; z_a^* = \frac{z_k}{L_k}\;, \text{respectively.}$$

Assuming the initial value for $T_{apk}(z_a^*= 0)$ vector, the spout region set of equations is

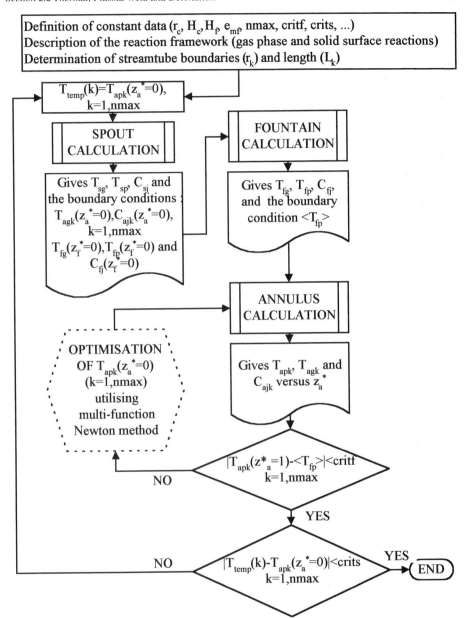

Figure 2 *Schematic diagram of the model resolution algorithm*

solved by the first call to STIFF3 package. Then, the values of temperatures and concentrations at $z = H_c$ are taken as initial conditions to solve the fountain system in a second call to STIFF3. Next step of the calculation is the resolution of the annulus equations. The values of particles and gas temperatures, of concentrations and conversions along the spout, obtained from spout calculation, are taken as initial values for annulus variables. The annulus region system of equations has also a boundary condition at the end of the stream-lines, which has to be fitted. This condition is formulated from fountain variables. Since STIFF3 method accepts only formulations with initial conditions, a shooting method was performed, for adjusting T_{apk} ($z_a{}^* = 0$) vector values in order to satisfy the boundary condition. The adjusted new values are calculated by a multi-variable Newton algorithm. If $T_{apk}(z_a{}^* = 0)$ values change significantly, the resolution of all the spouted bed reactor is performed again, beginning with spout calculation. The simplified schematic diagram of the algorithm is given in Figure 2 which shows the different steps in the model resolution.

3 RESULTS

A numerical study of the model has been performed. It has also been validated by comparison with Hook et al's[5] experimental data concerning the exothermic catalytic oxidation of CO over Co_3O_4/α-Al_2O_3 catalyst. In the present study, chemical vapour deposition (CVD) of silicon by monosilane pyrolysis is considered as an example of the model results.

Caussat et al[19] reviewed the published experimental and theoretical studies of this reaction in a fluidised bed at atmospheric pressure. The main possibilities of industrial applications of this process are silicon production by particle growth, and thin silicon film deposition on particles or immersed objects in the bed for oxidation or corrosion protection. The feasibility of this process has been demonstrated but two significant experimental limiting problems are reported: the parasitic fines formation and the particle agglomeration when the silane initial concentration exceeds a critical value.

All authors agree on the fact that silane pyrolysis in a fluidised bed leads simultaneously to heterogeneous silicon CVD and to homogeneous silicon decomposition. The heterogeneous decomposition of monosilane at atmospheric pressure is supposed to occur by the following overall reaction:

$$SiH_4 \rightarrow Si + 2\,H_2$$

Only a few kinetic equations can be found in literature, and the expression developed by Furusawa et al[20] is the most commonly used because it is specific for fluidised bed and it takes into account the inhibition effect of hydrogen. The first and the second terms on the right hand side represent respectively the heterogeneous and homogeneous reaction rates:

$$-\frac{dC_{SiH_4}}{dt} = \left(\frac{k_{so}\,S/V}{1 + K_{H_2}RTC_{H_2} + K_{SiH_4}RTC_{SiH_4}} + \frac{k_{vo}}{1 + K_u RTC_{H_2}} \right) C_{SiH_4}$$

where,

$$k_{so} = 2.15 \times 10^8 \exp(-191500 / RT)$$
$$K_{H_2} = 0.034 \times 10^{-3}$$
$$K_{SiH_4} = 7.6 \times 10^{-6} \exp(-32900 / RT)$$
$$k_{vo} = 2.14 \times 10^{13} \exp(-221300 / RT)$$
$$K_u = 0.5 \times 10^{-3}$$
$$S/V = \frac{6}{d_p} \frac{1-\varepsilon}{\varepsilon}$$

Caussat et al[19,21] recently developed an experimental analysis of a silicon CVD fluidised bed reactor and its modelling. We took several operating conditions among theirs for the SBR modelling: monocrystalline alumina particles are used as seed particles; nitrogen is the gas used; a low inlet silane concentration is chosen (to avoid reported particle agglomeration and because the model does not take into account variations of the volumetric flow-rate resulting from chemical conversion); the bed is heated through the reactor walls from the top of the cone to the top of the bed. Table 1 summaries the main input values of the SBR modelling.

Figure 3 shows the temperature and conversion profiles for the spout and the fountain regions versus the vertical dimensionless coordinate. The particle temperature profile in the

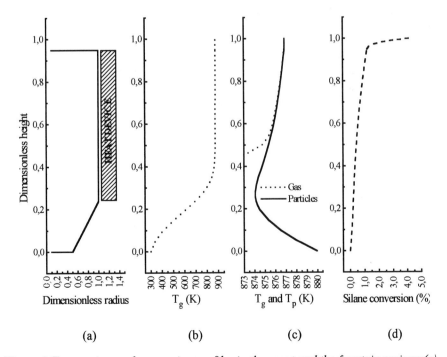

Figure 3 *Temperature and conversion profiles in the spout and the fountain regions: (a) reactor configuration, (b) gas temperature, (c) gas scaled and particle temperature, (d) silane conversion*

spout presents a minimum. It decreases in the first quarter of spout length because of the heat transfer with the entering cold gas. Then, as the gas temperature reaches that of the particles, the latter are increased because of the cross flow of hotter particles coming from the annulus. The overall particle temperature variation in the spout is low, about 6°C.

The fountain is quite isothermal for both solid and gas phases. The SiH_4 conversion at the top of the spout is 1.2%, it increases rapidly in the fountain to reach 4.4%.

The corresponding profiles of temperatures and conversion for the annulus along the stream-line dimensionless coordinate are plotted in Figure 4. The gas reaches the particle temperature just after entering the annulus. As the particles flow downwards, they are heated by heat transfer through the reactor wall, whereas in the cone zone, no energy is transmitted to the bed. The particle temperature stabilises or decreases because of the heat transfer with the colder gas from the spout.

In the annulus, the conversion variation follows the expected shape of a plug flow reactor (stream-tube) as a function of residence time. Streamtubes near the top of the annulus, having a smaller length, reach the smallest conversion levels. The global silane conversion in the annulus is 89%. Thus, as expected, the reaction mainly takes place in this region of the reactor. The overall reactor conversion is 75%.

4 CONCLUSION

A new non-isothermal model of a spouted bed reactor has been presented. It is based on simple hydrodynamic hypotheses allowing flow-rate calculation through known correlations;

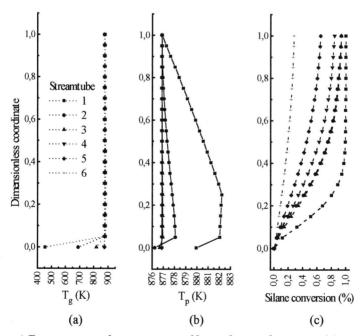

Figure 4 *Temperature and conversion profiles in the annulus region: (a) gas temperature, (b) particles temperature, (c) silane conversion*

Table 1 *Input parameters for reactor modelling*

Bed height	0.75 m
Cone height	0.225 m
Column diameter	0.15 m
Inlet tubing diameter	0.015 m
Flat bottom diameter	0.075 m
Particle diameter	0.001 m
Operating pressure	1 atm
Inlet temperature	300 K
Inlet mass flowrate	0.0197 kg s^{-1}
Inlet reactant mass fraction	0.05
Particle heat capacity	4795 10^3 J m^{-3}K^{-1}
Particle density	3965 kg m^{-3}

heat transfer is included by means of an heterogeneous approach. Silicon deposition from silane pyrolysis has been chosen to give an example of the model resolution. The reaction mainly takes place in the annulus region and a reasonable overall conversion is achieved in the reactor. The particle temperature varies in a narrow domain all over the reactor. The gas is heated very quickly up to the particle temperature.

The model will be improved by including the volumetric gas flow-rate changes resulting from chemical conversion. This will allow higher reactant concentration at the inlet, since diluted reactant will not anymore be a required hypothesis. Then, it will be possible to compare our SBR with other systems, and to optimise our operating conditions for best efficiency.

Notations

A	Section (m)
C_p	Calorific capacity (J m^{-3} K)
D	Diameter (m)
d_p	Particle diameter (m)
g	Acceleration due to gravity (m s^{-2})
H_c	Bed height (m)
H_f	Fountain height (m)
h_{lk}	Conductive heat transfer coefficient between adjacent streamlines (W m^{-2} K^{-1})
H_m	Maximum spouting height (m)
h_{pg}	Heat exchange coefficient between solid and gas phases (W m^{-2} K^{-1})
ΔH_{ri}	Heat of i reaction (J mol^{-1})
H_t	Total height ($H_t = H_c + H_f$) (m)
k_{vo}	Kinetic constant (s^{-1})
k_u	Kinetic constant (Pa^{-1})
k_{so}	Kinetic constant (ms^{-1})
k_{H2}	Kinetic constant (Pa^{-1})
k_{SiH4}	Kinetic constant (Pa^{-1})
L_k	k streamline length (m)
nmax	Total number of stream-tubes (-)
nreac	Number of independent chemical reactions (-)

Q_e	Energy transmitted to the bed through the reactor wall (W)
r	Radius (radial coordinate) (m)
r_i	Chemical reaction rate of i reaction (mol s^{-1} m^{-3})
R	Perfect gas constant = 8.314 (m^3 Pa mol^{-1} K^{-1})
T	Temperature (K)
U	Superficial gas velocity (m s^{-1})
V	Superficial particle velocity (m s^{-1})
W	Mass flow-rate (kg s^{-1})
z	Vertical coordinate (m)
z_k	Streamline integration direction (m)
z^*_s	($= z/H_c$), dimensionless coordinate in the spout (-)
z^*_f	($= z/H_p$), dimensionless coordinate in the fountain (-)
z^*_a	($= z_k/L_k$), dimensionless coordinate in the annulus (-)

Greek symbol

$\alpha(i,j)$	Stoichiometric coefficient of j specie in i reaction (-)
χ	Molar extent of reaction (mol m-3)
ε	Voidage (-)
λ	Thermal conductivity (W m^{-1} K^{-1})
μg	Gas viscosity (Kg m^{-1} s^{-1})
ρ	Density (Kg m^{-3})

Subscript

a	annulus
c	bed
f	fountain
g	gas
i	reaction
in	inlet
k	streamtube (k=1, reactor wall)
mf	minimum fluidisation conditions
p	particle
r	radial
s	spout

References

1. K. B. Mathur and C. J. Lim, *Chem. Eng. Sci.*, 1974, **29**, 789.
2. N. Piccinini, J. R. Grace and K. B. Mathur, *Chem. Eng. Sci.*, 1979, **34**, 1257.
3. K. Smith, Y. Arkun and H. Littman, *Chem. Eng. Sci.*, 1982, **37**, 567.
4. C. J. Lim and K. B. Mathur, *AIChE J.*, 1976, **22**, 674.
5. B. D. Hook, H. Littman and M. H. Morgan, *The Canadian J. of Chem. Eng.*, 1992, **70**, 966.
6. G. Rovero, N. Piccinini and A. Lupo, *Entropie*, 1985 , **124**, 43.
7. B. Thorley, J. B. Saunby, K. B. Mathur and G. L. Osberg, *The Canadian J. of Chem. Eng.*, 1959, 184.
8. C. J. Lim and K. B. Mathur, 1978, *Fluidisation* Proceedings of the 2nd Eng. Found. Conf. On Fluidization; J. F. Davidson and D. L. Keairns, eds, Cambridge University Press, London, p. 104.

9. T. Mamuro and H. Hattori, *J. Chem. Eng. Japan*, 1968, **1**, 1.
10. H. Littman, P. V. Narayan, A. H. Tomlins and M. L. Friedman, *AIChE Symp. Ser.*, 1981, **77**, 174.
11. G. A. Lefroy and J. F. Davidson, T*rans. Inst. Chem. Eng.*, 1969, **47**, T120.
12. D. Handley and P. J. Heggs, 1968, *Trans. Inst. Chem. Eng.*, **46**, T251.
13. P. N. Rowe and K. T. Claxton, *Trans. Inst. Chem. Eng.*, 1965, **43**, T321.
14. G. D. Mazza and G. F. Barreto, P*owder Technol.*, 1993, **75**, 173.
15. D. Kunii and J. M. Smith, *AIChE J.*, 1960, **6**, 71.
16. G. D. Mazza and G. F. Barreto, P*owder Technol.*, 1991, **67**, 137.
17. M. L. Michelsen, *AIChE J.*, 1976, **22** , 594.
18. G. F. Barreto and G. D. Mazza, *Computers Chem. Engng.*, 1989, **13**, 967.
19. B. Caussat, M. Hemati and J. P. Couderc, *Chem. Eng. Sc.*, 1995, **50**, 3615.
20. T. Furusawa, T. Kojima and H. Hiroha, *Chem. Eng. Sc.*, 1988, **43**, 2037.
21. B. Caussat, M. Hemati and J. P. Couderc, *Chem. Eng. Sc.*, 1995, **50**, 3625.

2.2.2

Surface Engineering by Boron and Boron-Containing Compounds

G. Kariofillis, C. Salpistis, H. Stapountzis, J. Flitris and D. Tsipas

ARISTOTLE UNIVERSITY OF THESSALONIKI, DEPARTMENT OF MECHANICAL ENGINEERING, LABORATORY OF PHYSICAL METALLURGY, THESSALONIKI, 54006, GREECE

1 INTRODUCTION

The surface properties of tools and dies used in materials manufacturing normally deteriorate with processing time. This deterioration is due to the presence of either aggressive environment, or the adverse operating conditons. One way to increase the life-time and performance of these components is to use coatings.

The coatings protect the substrate from excessive wear, erosion and corrosion and also improve the quality of the final products. Amongst the coatings that are extensively used for the protection of hot working tool steels are those of nitriding, boriding and electroless Ni-P. H13 tool steel is a steel commonly used as a die material for hot working operations, such as Al extrusion. Although various attempts have been made to protect tool and die steels, during hot working operations, with complex duplex coatings using for example combinations of plasma nitriding/PVD, the traditional treatment of gas nitriding remains amongst one of the most popular surface treatments for protecting dies used in Al extrusion. This is basically due to various factors including capital and operating costs, simplicity of the process, as well as good overall performance of the coating. Simple nitride, boride, or electroless Ni layers can also serve as a substrate for duplex layers of the type electroless Ni-P/borides, Ni-P/TiN and boriding/TiB$_2$, etc.

The diffusion of elements such as N, B and other small atoms into the surface of steels in order to obtain diffusion layers is carried out using gaseous, liquid, or solid substances and the basic chemical and physical properties of these simple diffusion layers and similar ones obtained by electroless deposition method have been reviewed[1-7].

In particular boride layers can be formed by:

- Immersion into hot saturated solutions containing boron compounds e.g. boric acid in alcohol.
- Exposure to vapours of saturated solutions containing boron compounds.

Evaporation can be done by thermal means, by ultrasonic stimulation of solutions, or nebulization and spraying. The important parameters for the immersion process are: the nature of the solvent, solute and temperature of immersion. For the exposure to vapours which is a process comparable to CVD the nature of vapours, the temperature and, the size of droplets during nebulization or ultrasonic stimulation play a critical rôle in the quality of coating.

In this paper we present the experimental results of a comparative study of the wear and high temperature erosion properties of nitride, boride, electroless Ni-P layers on H13 steel.

Attention is also given to the optimum heat treatment of H13 tool steel once the boride coating has been applied.

2 EXPERIMENTAL

The coating treatments were carried out applying the pack cementation method for boriding, the gas nitriding thermochemical method for nitriding, and a sodium chloride/sodium hypophosphite/lactic acid solution, for electroless Ni-P deposition. The heat treatment of H13 steel involved heating it at 1050°C followed by air cooling and tempering at 590°C. Coated boride specimens were heat treated at 850°C for 2 hours followed by water quenching and tempering at 600°C.

Coating characterization was carried out using optical electron microscopy and X-rays. The coating properties evaluation included microhardness, wear and erosion evaluation. Wear testing was carried out on a self-constructed equipment based on the pin-on-disk model. Details of the experimental set up are given elsewhere[4]. A schematic diagram of the erosion testing equipment is given Figure 11.

The equipment is basically a sand blasting ring. Test specimens are exposed to a known weight of an erodent, under pre-determined test conditions and then weighed to determine the weight loss. The particle feed control (1) determines the particle (erodent) flow. The particles enter chamber (2), while compressed air enters through inlet (9) and the particle feeder (3). This action creates suction which causes the acceleration of particles from chamber (2) towards particle feeder (3). The resulting air-particle mixture is directed towards the specimen holder (5) which supports the specimen at a given inclination (impact angle) and at a given temperature.

The particles impinge on the target (specimen) and fall by gravity to the particle collector (6). Any particles which do not fall and are carried by the air are trapped by the particle-fluid separator (7). When it is desired to introduce a corrosion medium, the corrosion medium feeder (8) serves to vaporise the medium e.g. NaCl solution before being injected into the particle stream. In this way the specimens can be exposed simultaneously to both corrosion and erosion conditions.

Erosion testing was carried out at three different temperatures 20°C, 200°C and 400°C using a gas flow rate of $Q_g = 10$ m³/h, gas velocity $V_g = 67$ m/sec, particle supply rate $Q_p = 1.85$ g/s, SiC particle average size of 300 μm, distance between sample and nozzle L= 50mm, and angle of incidence varied between 30-90°. The temperature was measured by placing a thermocouple in contact with the back side of the sample.

3 RESULTS AND DISCUSSION

The typical morphological characteristics and thickness of the boride, nitride and electroless Ni–P coatings are given in Figures 1–4. Nitride layer thickness was = 100μm, boride layers varied between 50–120μm depending on temperature, time of treatment and powder mixture used. The boride layers formed were either FeB/Fe$_2$B or single Fe$_2$B type, again depending on processing conditions. The electroless Ni-P thickness was around 10 μm with a high P content

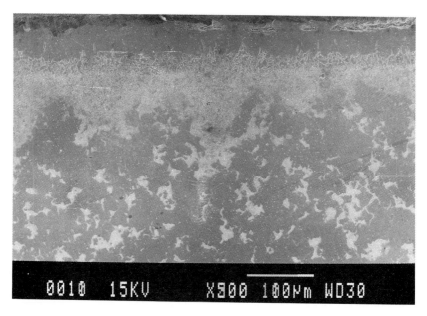

Figure 1 *Boride H13 (950 °C, 6 hours. There are two borides: FeB outer, Fe₂B inner, SEM*

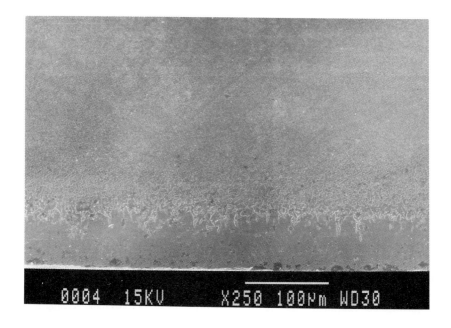

Figure 2 *Borided H13 with heat treatment (SEM)*

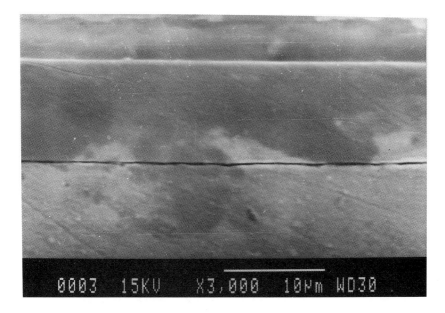

Figure 3 *Electroless NiP on H13 (Ni-9, 6%P, SEM)*

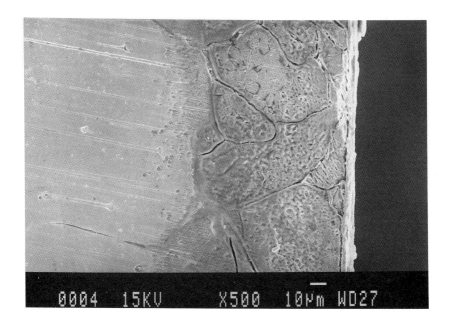

Figure 4 *Nitrided H13 (SEM)*

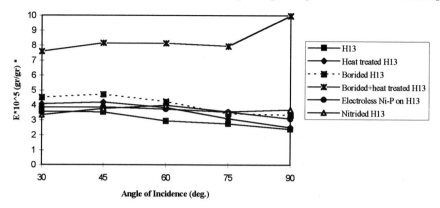

Figure 5 *Comparative data of erosion testing at different temperatures for various coatings on H13 steel (at 20°C)*

(P concentration = 9.69 wt.%). For the wear testing the Ni-P coatings were heat treated at 600°C for 1 hour.

The hardness distribution for the two coatings, the boride and electroless Ni-P are given in Figures 9 and 10.

The erosion results for the various coatings and testing conditions are shown in Figures 5–7. From these results it can be deduced that the lowest erosion rate was observed with H13 steel and the highest with borided and heat treated H13 steel. This is due to the better fracture toughness properties of H13, compared to the hard layers of borided steel. The heat treated electroless Ni-P and nitride coatings had very similar behaviour and no significant difference to the heat treated H13 steel for all tested temperatures. This result compared to the hardness values of these two coatings seems to be reasonable. For all practical purposes the erosion

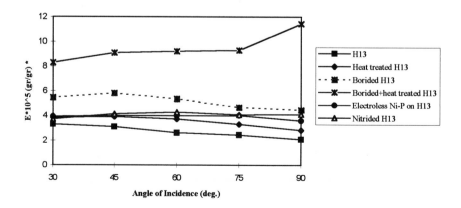

Figure 6 *Comparative data of erosion testing at different temperatures for various coatings on H13 steel (at 200°C)*

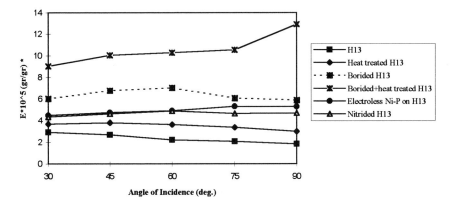

Figure 7 *Comparative data of erosion testing at different temperatures for various coatings on H13 steel (at 400°C)*

properties of these two coatings could also be considered as equivalent. This was not the case for the boride layers. The boride layers showed at all temperatures the highest erosion rate at high incidence angle confirming their brittle nature. This behaviour was observed for the first 10 minutes of erosion testing. After 10 minutes the weight change due to erosion seemed to remain unchanged indicating brittle-ductile mode of behaviour. More ductile materials such as heat treated and as-received H13 steel have the highest weight loss due to erosion at anges of incidence of 30° and 45°.

The wear testing results clearly indicated the boride coating had the best performance

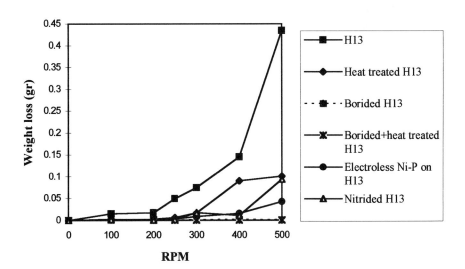

Figure 8 *Comparative data of wear behaviour of various hard coatings on H13 steel (weight 500g)*

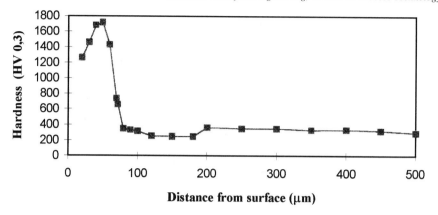

Figure 9 *Borided H13 (950ºC, 6 hours) with heat treatment*

from all coatings tested while Nitride and Ni-P coatings again presented very similar properties.

From the above results it is clearly seen that from the tested coatings none possess the optimum properties for both erosion and wear conditions. An interesting duplex coating combination would be a boride coating on top of electroless Ni-P.

4. CONCLUSIONS

Boride coatings showed the best wear properties and the highest erosion rate – Ni-P and nitride layers had very similar wear and erosion behaviour.

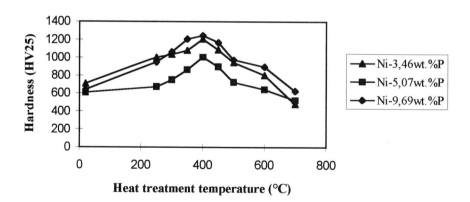

Figure 10 *The relationship between temperature and hardness of electroless Ni-P (acid solutions) on H13 steel*

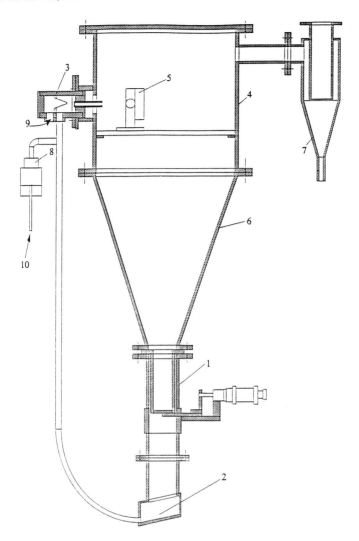

1. Particle feed control
2. Particle chamber
3. Particle feeder
4. Reaction chamber
5. Specimen holder
6. Particle collector
7. Particle-fluid separator
8. Corrosion medium feeder
9. Main compressed air inlet
10. Secondary compressed air inlet for corrosion medium transport

Figure 11 *Corrosion-erosion equipment*

References

1. G. A. Dearnaley and T. Bell, 1985, *Surf. Eng.*, **1**, 203.
2. A. G. von Matuschka, Boronizing, Hanser and Heyden, 1980, Berlin.
3. D. Tsipas, J. Russ, Surface Engineering Practice, Edit. S.K. Strafford, P.K. Datta, J.S. Gray, Ellis Norwood Publishers, 1990, p. 102.
4. A. S. M. Kurney, M. Mohan Rao and R. M. Mallya, Tool and Alloy Steels, Alloy Steel Producers Association of India,1983, p. 333.
5. F. Pearlstein, in Electroplating, F.A. Lowenhein, pub. McGraw-Hill Book Company, 1993.
6. L. G. Yu, X. S. Zhang, *Thin Solid Films*, 1993, **220** , 76.
7. L. Shoufu, M. Erming, L. Pengxing, 1986, *J. Vac. Sci. Technol.*, A4, Nov/Dec 1986.
8. S. Skolianos , T. Z. Kattamis, *Mater. Sci. Eng.*, 1993, **A163**, 107.

2.2.3
Study on Carburizing/Boriding Compound Chemical Heat Treatment and its Application

P-Z. Wang[1], L-Y. Shan[1], Z-Y. Ni[1] and P. Ye[2]

[1]CHINA UNIVERSITY OF MINING AND TECHNOLOGY, CHINA

[2]CHINA COAL MINISTRY, CHINA

1 INTRODUCTION

Owing to the superior comprehensive properties, compound chemical heat treatment processes have been used more widely, replacing single treatments[1~7]. For example, boriding-sulphurizing compound processes have been applied to Cr12MoV steel moulds[8]; a hard boronized layer (Fe_2B or $Fe_2B + FeB$) was covered with a thin sulphurized coating (FeS, FeS_2). The porous structure is helpful in absorbing lubricant oil and the hcp atomic structure of iron sulphide itself acts as lubricant; in this way, scuffing and galling will be alleviated, meanwhile iron boride resists adhesion and abrasion. As for carburizing-boriding treatment, the carburizing layer supports the thin, hard and brittle boride layer, then the compound layer will be more resistant to fatigue wear and adhesive wear. This will enlarge the application of the boriding process. The present paper introduces the research work of carburizing-boriding compound treatment: the regularity of layer formation, the microstructure observation, the properties evaluation and application on tricone bit parts.

2 LAYER FORMING REGULARITY AND MICROSTRUCTURES

2.1 Experimental Purpose

It is known that layer forming regularity will be different for single treatment and compound treatment. Experiments show precarburizing increases the speed of subsequent boriding[9]. Since 20CrMnTi (Chinese YB) is used as a carburizing steel for bearings, gears and shafts, which work under fatigue condition, this paper uses 20CrMnTi as an experimental material. Experiments were arranged to reveal the forming regularity of compound layer in order to provide data for practical application.

2.2 Experimental Conditions

2.2.1 Experimental Material. 20CrMnTi was used as the substrate material, its composition is shown in Table 1.

2.2.2 Carburizing-Boriding Process. The carburizing equipment was a drip gas carburizing furnace. The carburizing agent was kerosene and methanol. Carburizing process 930°C, conditions were 3~4 h, air cooling; the case depth was 0.7~0.8 mm. Liquid boriding process was carried out, KXB-1 type boriding salt bath was used.

Table 1 *The composition of experimental steels, wt. %*

	C	Cr	Mn	Ti	Ni	Mo
20CrMnTi	0.21	1.20	1.10	0.09	0.09	---
20CrNiMo	0.21	0.90	0.50	---	1.20	0.20
15CrNi3	0.16	0.80	0.40	---	3.30	---
16Mn	0.16	---	1.40	---	---	(Si) 0.40

2.2.3 Boriding Experimental Results. Figure 1 and Figure 2 show the influence of heating temperature and time on boride depth of carburized and uncarburized specimens. By comparison it can be seen that precarburizing increased the subsequent boriding speeds; under the condition of this paper, the boride depth of the carburized specimen was 10.1% (average value) deeper than that of the uncarburized specimen after the same boriding process.

Figure 3 shows the microstructure of specimen treated by the carburizing-boriding process. Carbon and boron element distributions across the surface layer were revealed by wave spectrometer, as shown in Figure 4.

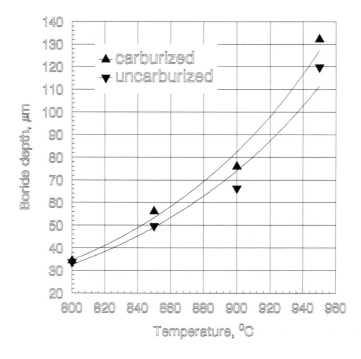

Figure 1 *The influence of temperature on boride depth of carburized and uncarburized specimens*

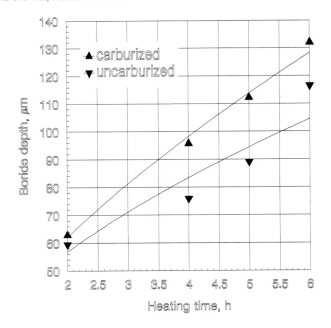

Figure 2 *The influence of time on boride depth for carburized and uncarburized specimens*

3 WEAR TESTS

3.1 Fatigue Wear Test

The experiments were carried out on a JP-BD1500 type contact fatigue tester. The specimens are shown in Figure 5. The specimen's material was 20CrMnTi, carburized and boronized. Carburizing

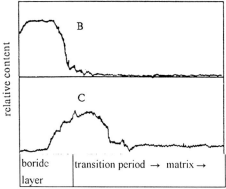

Figure 3 *The microstructure of carburized + boronized specimen*

Figure 4 *The element distribution across carburized + boronized layer*

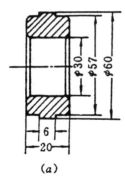   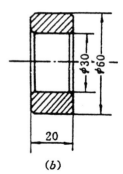

         (*a*)   (*b*)

Figure 5 *The specimens of fatigue wear tester, (a) upper specimen, (b) lower specimen*

case depth was 1.6~1.8 mm, boride depth was 80 μm and 120 μm. Totally circulating lubrication was facilitated by 20# engine oil. The running speed of the upper and lower rings Ur_u = 1000 and Ur_l = 800 rpm respectively, then the relative sliding ratio was 20%, i.e. 0.523m/s. According to primary experimental results and by adjusting load, the pressure, p_0, along the central line of the contacting area (by Hertz theory) is kept to 3200 MPa.

The fatigue wear resistance was compared by measuring the duration until delamination occurred to either the upper or the lower specimen. At that time vibration and noise increased suddenly, and the testing machine was stopped immediately.

One pair of carburized specimens was used as standard specimens, their service life was regarded as 1. The experimental results are shown in Table 2. It can be seen that carburizing + boriding process treated specimens had higher fatigue wear resistance.

3.2 Dry Sliding Wear Test

The equipment was a MM-200 (Amsler) wear tester, specimen size was φ 50 × 10 mm. The material was also 20CrMnTi again, treated by carburizing + boriding process. The case depth was 0.7~0.9 mm. Boride depths were 80,120 and 160 μm. The running speeds of the upper and lower specimens were 180 and 200 rpm respectively, the relative sliding speed was 0.052m/s. The load was L = 1000N, i.e. the pressure p_0 = 530 MPa. Experiments showed the wear debris was black powder, the wear mechanisms were adhesion, abrasion and oxidation. By measuring weight loss after a certain sliding distance, the wear resistance was compared. A pair of carburized specimens were used as standard specimens. The results are shown in Table 2.

Table 2 *Wear test results*

	Carburizing	Carburizing + boriding		
Boride depth, μm	0	80	120	160
Fatigue wear resistance	1	2.2	3.1	---
Dry sliding wear resistance	1	3.7	5.3	3.6

4 SIMULATED TEST

The simulation tester is manufactured by Shanghai No1 Petroleum Machinery Factory, which imitates the working conditions of tricone bits used in oil fields. The specimens are shown in Figure 6.

The shaft specimen material was (Chinese YB) 20CrNiMo, its composition is shown in Table 1. The shaft specimens were treated by carburizing + boriding process, carburizing case depth was 1.0 ~ 1.2mm, boriding depth was 80, 120 and 160 μm. The bearing specimen was made of (Chinese YB) 15CrNi3 steel, its composition is shown in Table 1. The bearing specimens were carburized, to a case depth 1.7~1.9 mm and then quenched.

The experimental conditions were established to simulate practical working condition of: drilling pressure 16~18 ton, drilling speed 60 rpm, under which if the tricone bit can work for 80 hours or more, they will be regarded as *first class products* according to the standards of the Chinese Machinery Ministry. Copper and Ag-Mn alloys were used as lubricant. Table 3 lists results for shaft

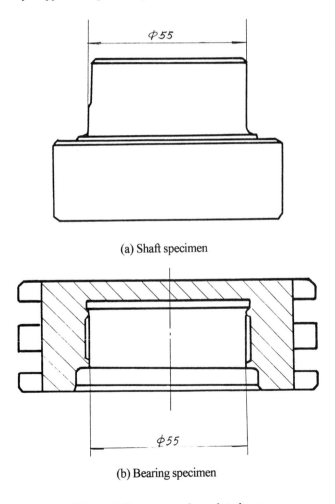

(a) Shaft specimen

(b) Bearing specimen

Figure 6 *Specimens of simulated test*

Table 3 *Observation of carburized + boronized shaft specimens after simulated test*

No	Boride depth (μm)	Observation of worn surface
1	80	Small and shallow delamination
2	80	No galling, no obvious delamination
3	120	Small and shallow delamination
4	120	No galling, no obvious delamination
5	160	No galling, no obvious delamination
6	160	No galling, no obvious delamination

specimens after 80 hours simulation.

Figure 7 shows a shaft specimen after simulated test. The experimental results meet the requirements for first class drills set by Chinese Machinery Ministry.

5 BENCH TEST

Nine sets of new tricone bits have been manufactured with their shafts treated by the carburizing-boriding process. They were divided into three groups according to the boride depth: 80, 120 and 160 μm. At the request of the manufacturing factory, the bench test was simplified as only 120 μm group bits were tested. The pressure and speed were the same as that of simulated test. Hot rolled plates of 16Mn (Chinese YB) steel were drilled, with water as lubricant. After 60 hours the bit still worked well, this also met the requirements of above mentioned first class products standards.

6 CONCLUSION

1. Under the experimental conditions of this paper, precarburizing increased the subsequent boriding speed, by an average of 13.9%.
2. Wear resistance of specimens treated by the carburizing-boriding process is 2.58 times (averaged) higher than that of carburized ones.
3. Simulated test and bench test showed the application of carburizing-boriding process to tricone bits is very promising.

Figure 7 *Shaft specimen after simulated test*

References

1.   H. Yizhen, *Chemical Heat Treatment Abroad*, 1982, No2, p. 4 (in Chinese).
2.   *Chemical Heat Treatment Abroad*, 1982, No1 p. 1 (translated into Chinese).
3.   S. Fujiki, *J. Jpn. Soc. Heat Treat.*, 1992. **32**, 303 (in Japanese).
4.   T. Lei and D. Sun, *Heat Treat. Met.* (China), Aug. 1991, 13 (in Chinese).
5.   A.V. Reddy, *Surf. Eng.*, 1992, **8**, 136.
6.   S. Cong and L. Qiu, *Heat Treat. Met.* (China), Jan. 1992, 54 (in Chinese).
7.   US Patent No.4188242.
8.   M. Ohyang , *Heat Treat. Met.* (China), May 1984 (in Chinese).
9.   P-Z. Wang, P. Ye, Z-Y. Ni and G-Z. Xie, *Proceedings of the 5th World Seminar on Heat Treatment and Surface Engineering, IFHT'95*, September 26–29 1995, Isfahan, Iran, 349.

2.2.4

An Integrated Study of the Structure and Durability of Vacuum Furnace Fused Coatings

T. Hodgkiess[1] and A. Neville[2]

[1]DEPARTMENT OF MECHANICAL ENGINEERING, GLASGOW UNIVERSITY, GLASGOW, SCOTLAND

[2]DEPARTMENT OF MECHANICAL AND CHEMICAL ENGINEERING, HERIOT-WATT UNIVERSITY, EDINBURGH, SCOTLAND

1 INTRODUCTION

There is evidence from a large range of industries (e.g. the oil and gas industry[1,2] pulp and paper manufacture, chemical and mineral processing[3–5] and mining applications[6–9]) that, in severe erosion-corrosion conditions, even the high-grade metallic materials (such as stainless steels) can suffer rapid rates of deterioration. This leads to economic penalties due to inefficient operation of equipment, reduced operational output and adverse environmental effects. To consider, as an example, the operation of rotodynamic equipment in an oilfield application, the requirements of increased output from drilling wells pose a severe problem in the operation of downhole equipment. To accommodate larger volume flow rates, the equipment must operate at much increased rotational velocities and with the added severity due to entrained solids, erosion-corrosion can lead to failure in pumps and other related equipment in relatively short service periods.

These problems have led to significant activity over the past two decades to enhance material performance through surface engineering. There is a wide range of chemical and physical treatments including chemical vapour deposition (CVD), nitriding, physical vapour deposition (PVD), shot-peening and gas carburising, but one of the most promising areas of surface engineering has been the development of thermal spray techniques, specifically for the application of advanced ceramic-base coatings. It is in the area of thermal spray coatings that the focus of this paper lies.

Although still accounting for a very small proportion of the total UK surface engineering market[10], thermal spray processes offer great potential as a means of combating premature aqueous erosion-corrosion failures. However, it is accepted within the industry that their full potential has not yet been realised and exploitation has been limited within particularly the oil and gas industry. The reasons for this are based mainly in the lack of reliable information on their performance characteristics in severe environments and hence a resulting reluctance by operators to 'experiment'.

Cermet and ceramic-base coatings have traditionally been developed to enhance material resistance to mechanical wear. However, it is common in a range of industries for these materials to be required to operate in a corrosive aqueous stream and as such, the resistance to corrosion is of as much importance as wear resistance. Hence the thermal sprayed coating is often required to act as a barrier to the ingress of a corrosive fluid to the substrate. Porosity of thermal sprayed coatings, whether applied by plasma techniques, high velocity oxy fuel (HVOF), plasma-assisted chemical vapour deposition (PACVD), detonation gun (D-Gun) or other related techniques has become the focus of several studies involving porosity determination and sealing techniques[11-13]. In this work the coating (applied by oxy-acetylene thermal spray method) and substrate are subjected to a vacuum annealing process which, by consolidating the coating constituents, is intended to remove the requirement for sealant application.

Assessment of the wear characteristics of thermal sprayed coatings has traditionally involved weight loss determination in a range of erosive-corrosive environments[14]. In addition, manufacturers have reported particular 'success stories' of their product based on service experience[15] A few recent studies have applied electrochemical techniques to static corrosion studies of metallic and cermet coating systems[16,17]. However there is an inherent need, in studies of complex multiphase systems of the type described here, to bring together information from multidisciplinary studies which will enable the understanding of the corrosion and erosion-corrosion processes. This will involve an integrated experimental approach involving aspects of electrochemical corrosion monitoring, comprehensive microscopy and microanalysis before and after exposure to a range of environments. Research must be focused towards assessing the global performance of a coating/substrate system and then progressing to look, on a micro-scale, to the interactions which exist between the coating constituents in order to predict the components most likely to suffer enhanced attack and hence limit the performance of the coating either due to corrosion, mechanical erosion or the joint processes.

As a means of illustrating some of the factors involved in overall durability assessments of wear resistant cermet coatings, this paper focuses on one type, i.e. thermally-sprayed, vacuum fused nickel-chromium-boron and cobalt-chromium-boron-tungsten, both coatings applied to a carbon steel substrate.

2 EXPERIMENTAL

Two basic varieties of thermally-sprayed, vacuum fused coatings were investigated: a cobalt-chromium-tungsten-boron-base cermet and a nickel-chromium-boron-base material. The Co-base coatings are primarily hardened by the presence of precipitated tungsten carbide with, in addition, dispersed chromium and nickel carbides and borides. In contrast, the Ni-base coatings consist of chromium carbides, borides and nitrides as the hard phase. Two grades of each coating type were employed to give a total of four coating materials:-

- cobalt-base with Rockwell C hardness of 40, hereafter referred to as 'Co/40'
- cobalt-base with Rockwell C hardness of 50, hereafter referred to as 'Co/50'
- nickel-base with Rockwell C hardness of 50, hereafter referred to as 'Ni/50'
- nickel-base with Rockwell C hardness of 60, hereafter referred to as 'Ni/60'

The compositions of the four coatings are given in Table 1.

The coatings were applied to a BS 970 EN8 carbon-steel plate by a specialised coating company. Specimens, 1×1 cm for corrosion and microscopical studies and 3×2 cm for erosion tests, were cut from the as-received plates. Electrical-connecting wires were soldered to the back (steel) face of the samples prior to encapsulation in castable resin. Finally, the samples were ground on abrasive papers and polished to a one-micron diamond finish.

Corrosion experiments were carried out by immersion of the specimens in a seawater solution made up using a proprietary product, 'Ocean Salt', which contains all the major ionic

Table 1 *Nominal compositions of the Co and Ni-base coatings*

	%Cr	%B	%Si	%Fe	%W	%C
Co/40	19.0	1.7	3		8	0.7
Co/50	17.5–20.5	2.25–2.75	2.5–3.5	3%max	12.5–14.5	1.2–1.5
Ni/50	10.5–12.5	2.25–2.6	3.25–4	3.2–4		0.45–0.55
Ni/60	13.5–15.5	2.85–3.15	4.4–4.75	4.4–4.75		0.55–0.8

constituents of seawater, dissolved in distilled water to yield a salinity of 35,000 ppm. The specimen/resin interfaces of the encapsulated specimens were painted with a sealing lacquer, 'Lacomit', and, 1 hour after being initially immersed in the seawater at ambient temperature (18°C), the specimens were subjected to standard DC anodic polarisation potentiodynamic scans carried out at 15 mV/min. All potentials quoted herein were measured using the saturated calomel electrode. Additionally, two sets of samples were exposed to the seawater, at ambient temperature, for extended periods of 64 days and 40 days with periodic monitoring of electrode potential and visual inspection. At the completion of these extended exposure periods, anodic polarisation scans were conducted on some specimens whilst others were removed from the water without any polarisation.

The other durability experiments comprised erosion-corrosion tests in a rig described elsewhere[18,19] in which a fine liquid jet, approximately 1mm diameter, impinged perpendicularly onto the sample surface at a velocity of 100 m/s with a jet-to-specimen distance of 1 cm. These experiments were carried out in a 3.5% (commercial grade) NaCl solution containing a very-small population (less than 10 mg/l) of solid particles. These tests were continued for 1 hour at the end of which some specimens were removed undisturbed whilst others were subjected to an in-situ anodic polarisation sweep before terminating the erosion-corrosion experiment.

The durability studies were supported by microscopical examination of as-received coatings (in plan and cross-section) and specimens after corrosion and erosion-corrosion tests employing light and scanning-electron microscopy (SEM). Microanalysis was undertaken using energy-dispersive equipment on the SEM and the facility also provided quantitative (atomic number, absorption and fluorescence) correction procedures for spot analyses.

3 RESULTS

3.1 Microscopical Examination of 'As-received' Coatings

3.1.1 Coating Structure. The cermet coatings have extremely-complex microstructures. These will be discussed by reference to the results of examination of one each of the two types of coating investigated in this work. The other two coatings differ essentially only in the amount of hard phase present.

Ni/50 Coating – Examination of polished specimens in plan on the light microscope revealed (Figure 1) a number of different phases. Two of the microstructural features could be seen to stand out in relief from the matrix thus identifying them as hard phases:-

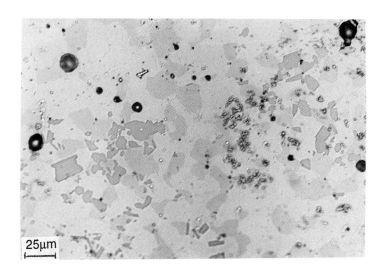

Figure 1 *Microstructure of Ni/50 coating as-received*

i.   large, irregularly-shaped particles (slightly darker than the matrix in Figure 1) and fairly
     uniformly distributed throughout the structure;
ii. bundles of irregularly-distributed, much-smaller, blue-grey globular particles visible on
     the right-hand side of Figure 1.

Additionally, there are darker, spherical particles usually with a lighter-coloured core.

The large, hard-phase particles, (i) above, are clearly evident (Figure 2) as the dark
constituent in backscattered SEM images and spot analyses on a number of these yielded a
chromium concentration of between 77–81.5% with 1.7–2.1%Ni, 0.3–1.3%Fe, 0.1%Si. These
clearly represent the major, chromium carbide hard phase introduced into this coating.

Analysis on the matrix gave the expected high nickel content (around 85%) with 5–7.5%Cr
and around 4% Fe and Si.

Accurate analysis on the spherical, dark particles in Figure 2 was difficult due to their small
size but they were found to be relatively rich in silicon (28%Si measured).

The large grey particles in Figure 2 can be seen to be of similar shape and size to the darker
ones and spot analyses gave similar chromium contents on these two features. Nickel analysis
on three different lighter-grey particles yielded 2.5, 8.8, and 8.9%Ni and these are considered
to represent the same constituent as the darker grey constituent but sitting just below the
polished surface.

An additional feature observed on the SEM comprised needle-like particles. Accurate
analysis on these was impossible because of their size (about 1 micron) but the acicular particle
in the centre of Figure 3 gave 44.1%Cr – a good indication that these are chromium rich and
are presumed to be an acicular chromium carbide phase.

Co/40 Coating – Examination of polished specimens in plan on the light microscope revealed
a uniformly-distributed network of the hard phase in relief in Figure 4. Another major constituent

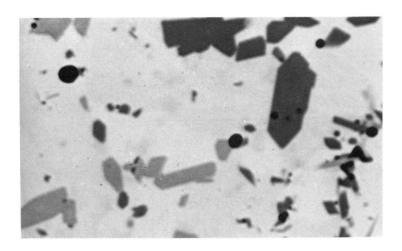

Figure 2 *SEM image of Ni/50 as-received showing hard particles (dark phase)*

comprises globular black particles which, on closer examination, Figure 5, can be seen to possess a slightly lighter core. Also just evident in Figure 5 are: (i) tiny acicular particles and (ii) minute spherical particles.

The backscattered SEM images (Figure 6) reveal a dense distribution of bright (i.e. high atomic number) particles. These and the other major constituent – the black particles – observable in Figure 6 are shown at higher magnification in Figure 7. Point analyses on the features visible in Figure 7 are shown in Table 2.

Other analyses on the four features yielded very similar elemental concentrations to these tabulated above. It is clear that the white and black particles are the tungsten-rich and chromium-rich hard phases respectively. The detection of significant concentrations of other elements in these two phases probably arises from interdiffusion between the hard phases and the Co-Cr-base matrix during the vacuum annealing treatment.

Another feature evident in Figures 6 and 7 and shown in more detail in Figure 8 is a duplex structure typical of a eutectic. The constituents of this appear to be the coating matrix and the grey phase shown in Figure 7 and analysis 'B'. The as-received Co-base coating also showed evidence of circular Si-rich particles similar to those described above in the Ni-base coating.

Table 2 *Composition of phases in Co/40 coating as determined by EPMA*

Feature	%W	%Co	%Cr	%Ni	%Fe	%Si
'A', bright particle	53.9	26.3	12.9	3.1	0.5	2.7
'B' ,grey particle	9.9	24.8	47.7	3.9	1.8	1.9
'C', matrix	0.7	57.8	12.7	18.3	3.9	4.3
Black particle	3.4	20.0	62.4	1.6	1.4	0.3

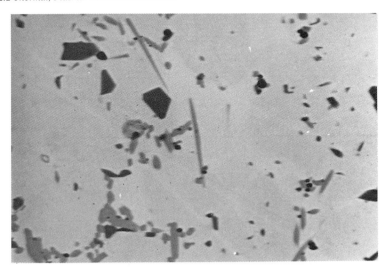

Figure 3 *SEM image showing needle-like particles on Ni/50, as-received*

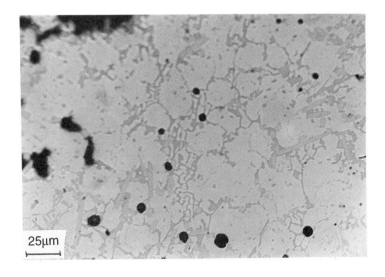

Figure 4 *Microstructure of Co/40 coating as-received*

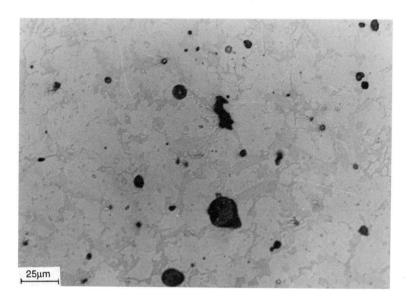

Figure 5 *Black spherical particles with a lighter core in Co/40 coating, as-received*

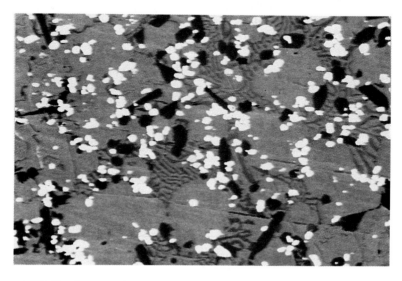

Figure 6 *SEM backscattered image Co/40 coating, as-received, showing distribution of high atomic number (light) particles*

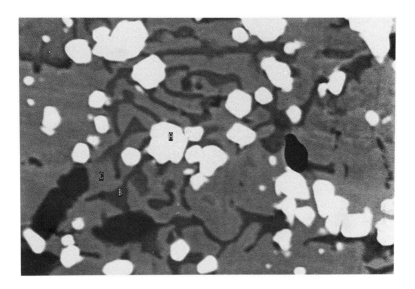

Figure 7 *Areas of spot analyses on Co/40 coating, as-received*

The above description has related to the structure of the two types of coating as examined in plan. The structure in depth is obviously important and examination of polished cross-sections revealed a generally uniform structure with good coating/substrate adhesion as illustrated for a Co/40 coating in Figure 9. There was evidence of some residual porosity in the cross-section examination of the coatings. Also, as shown in Figure 9, and at higher magnification on a Ni/50 coating in Figure 10, there was a thin, featureless layer at the coating/substrate interface. Analysis on this layer in Figure 10, indicated a somewhat lower nickel content (about 78%) and higher iron concentration (about 10%) than in the bulk coating. This is due to interdiffusion between the coating and steel substrate during the vacuum annealing operation. The distribution of the constituent elements is demonstrated in the X-ray images for two of the coatings shown in Figures 11 and 12. One feature of interest is the distribution of Si-rich particles, mentioned previously, which can be seen in Figure 11a.

3.2 Corrosion Tests

During the anodic polarisation scan after the initial one-hour exposure to seawater, specimens of all four coatings exhibited (Figures 13 and 14) low currents over a range of potential positive to the free-corrosion potential, E_{corr}. This is typical of a low corrosion rate afforded, for instance, by the presence of a passive film on the surface. At a sufficiently-high potential (denoted E_b), the recorded currents increased rapidly signifying the onset of rapid corrosion. As indicated by the more-positive values of E_b, the nickel-base coatings exhibited somewhat greater resistance to breakdown of corrosion protection than the cobalt-base coatings. It was consistently observed that the coating (whether Co- or Ni-based) of higher Rockwell hardness exhibited a slightly lower (approximately 50mV) value of E_b. For example, Figure 14 indicates

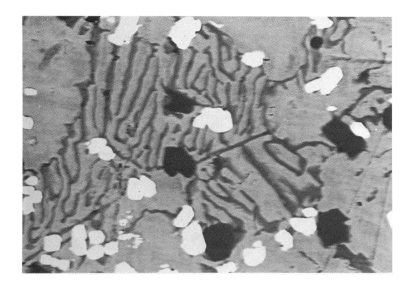

Figure 8 *Duplex, eutectic-like structure on Co/40 coating, as-received*

Figure 9 *Cross section of the Co/40 coating showing good adhesion*

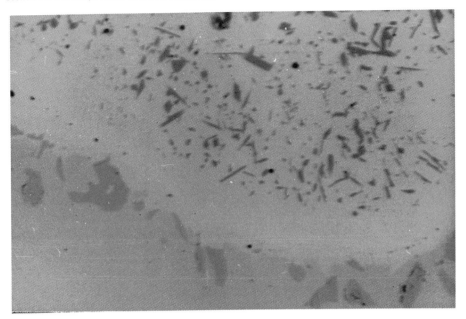

Figure 10 *Featureless layer at the base of the coating of Ni/50, adjacent to the substrate*

that the Co/40 coating is more resistant to the onset of severe corrosion than the Co/50 material.

The anodic polarisation sweeps were programmed to continue until a current of 500 microamps at which point the potential scan was reversed. For these one hour exposures, the potential reversal was accompanied, on all four coatings, by an almost immediate decrease in the anodic current but the wide hysteresis loop (Figures 13 and 14) indicated significant difficulty in the re-establishment of the low corrosion rates typical of the forward scan situation.

Although just one set of results are shown in Figures 13 and 14, three sets of repeated experiments (employing different batches of material) yielded good agreement with those shown herein.

The corrosion behaviour after longer-term exposure in seawater was found to be quite complex. Thus, anodic polarisation tests, undertaken after an exposure period of 40 days, were of a similar form to those after one hour but the currents recorded in the potential range immediately positive to E_{corr} were observed to be greater than on the one-hour samples and also, upon potential-scan reversal, the current continued to increase up to a value, for example, of 1000 microamp on the Ni/60 coating. These features are shown for one Co- and one Ni-base coating in Figure 15. Also, there were clear visual signs of crevice corrosion on the specimens. In the first long-term exposure experiment, the crevice corrosion could be seen (Figure 16) to emanate from the lacomit film at the specimen/resin interface. Although it was considered that this corrosion represented crevice attack at the lacomit/cermet coating interface, the possibility was recognised that the corrosion emanated from the bare steel face where it intersected the surface. Thus another set of specimens was prepared with the normal lacomit seal at the specimen/resin interface but additionally a strip of lacomit across the centre of the coating. These samples were immersed in seawater and, after a period of 20 days, evidence

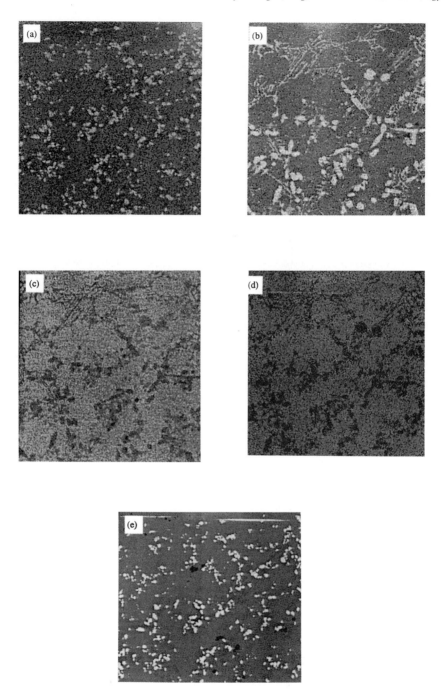

Figure 11 *X-ray images showing elemental distribution in Co/50 coating: (a) W, (b) Cr, (c) Co, (d) Ni and (e) backscattered image (BSE)*

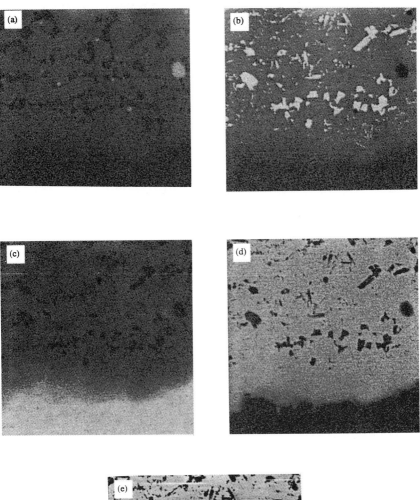

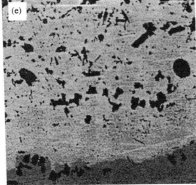

Figure 12 *X-ray images showing elemental distribution in Ni/50 coating: (a) Si, (b) Cr, (c) Fe, (d) Ni and (e) backscattered image (BSE)*

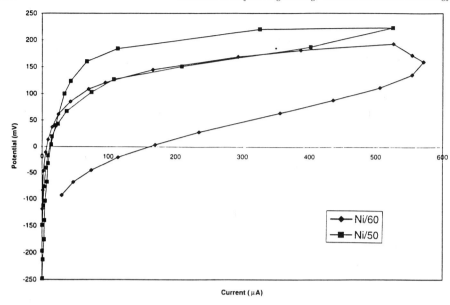

Figure 13 *Anodic polarisation on the Ni-based coatings after 1 hour immersion in seawater*

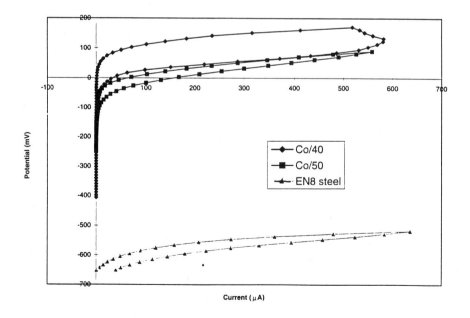

Figure 14 *Anodic polarisation on the Co-based coatings after 1 hour immersion in seawater and on the EN 8 substrate*

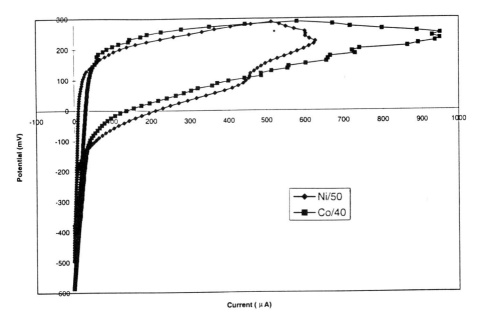

Figure 15 *Anodic polarisation on the Co/40 and Ni/50 coatings after 40 days immersion in seawater*

(Figure 17a) of corrosion initiating at the lacomit/coating interface in the centre of the coating was observed on the Ni-base material. At this time, there was no sign of corrosion on the Co-base samples (see Figure 17b) but, in the subsequent period, similar crevice corrosion at the central lacomit strip initiated so that, by the end of the experiment (40 days), all four types of coating were affected. During this exposure period, free corrosion potentials were monitored and there were shifts in E_{corr} observed but with no systematic relation to the onset of crevice attack.

3.3 Erosion-Corrosion Experiments

At the completion of the one hour impingement tests, all specimens (comprising all four coating types) were found to have undergone damage in the small zone directly underneath the impinging jet – Figure 18 shows this effect. It was not possible to obtain any quantitative comparison of the relative degree of damage experienced by the four different materials; this requires a more extensive programme which is planned. However, some comparison was secured from the anodic polarisation tests carried out on two of the materials during the final stages of the jet impingement test. Thus duplicate specimens of the Co/50 coating exhibited (Figure 19) active corrosion behaviour, i.e. rapidly increasing current upon commencing the potential sweep positive from E_{corr}. In contrast, 'in-situ' anodic polarisation on the Ni-40 specimen showed initially a potential range of steadily increasing current (Figure 20) larger than the analogous currents observed in static conditions after a one hour exposure. Subsequently, a large increase in current, similar to that observed at E_b in static conditions, was recorded. This behaviour can best be interpreted as the manifestation of severe active corrosion at and just above E_{corr} in the zone directly under the jet with the remainder of the

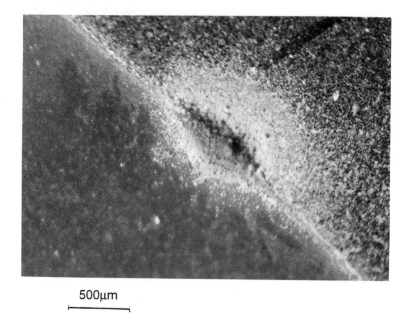

500μm

Figure 16 *Crevice corrosion at the Iacomit/coating interface on Co/40*

specimen surface exhibiting substantially lower corrosion rates. Subsequently when the potential reached E_b, the increase in current was associated with the onset of corrosion over the entire specimen surface.

3.4 Post-Test Examination

As mentioned previously, the Co- and Ni-base coatings in this study are multi-phase, comprising an array of hard particles embedded in a metallic matrix. As expected, post-test examination of corrosion and erosion-corrosion specimens confirmed that such complex structural features are related to material degradation mechanisms.

This study has demonstrated that, in saline environments, the coatings are susceptible to both general surface attack and localised corrosion under free corrosion conditions. The most severe type of attack was found to be crevice corrosion, which, on the Ni-base coatings, occurred at the sealing lacquer/coating interface after relatively short periods of 20 days.

After 40 days immersion, severe attack was evident on the Ni-base coatings and less severe attack on the Co-base coatings. Microscopic examination of this crevice corrosion revealed complex surface characteristics; almost the entire surface, with the exception of the hard phase particles, appeared to have been attacked. The attack was characteristic of intergranular corrosion as shown by a region of initiation in Figure 21 which resulted in a skeletal structure of hard phase remaining with large deep areas of matrix removed. It is postulated that the propagation of intergranular attack led to entire matrix grains being removed (Figure 22).

Additionally there was attack on the free surfaces of the coatings. This involved both pitting and general corrosion. In the form of pitting, clear attack was found on the free surface

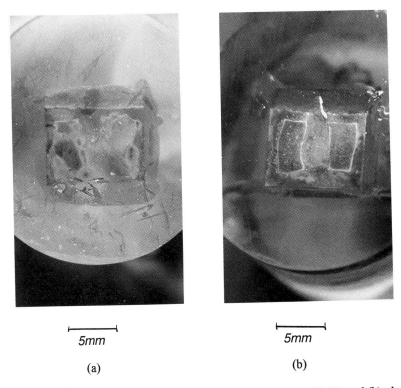

5mm

(a)

5mm

(b)

Figure 17 *(a) Crevice corrosion at the lacomit/coating interface on Ni/50 and (b) almost corrosion-free interface on Co/40. Both specimens examined after 20 days immersion*

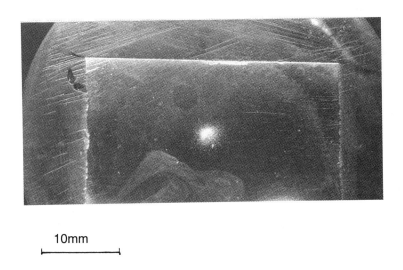

10mm

Figure 18 *Depression on Ni/50 after 1 hour exposure to the impinging jet at 100m/s*

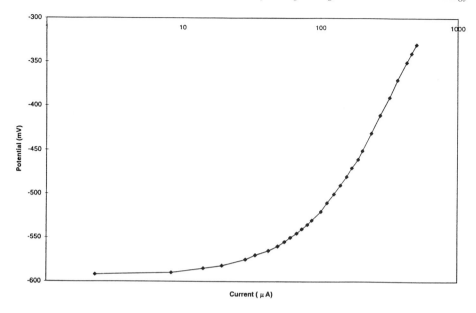

Figure 19 *Active corrosion behaviour on Co/50 under the impinging seawater jet*

of both the Co- and Ni-base coatings as shown in Figure 23. Some evidence was obtained to suggest that this pitting was associated with the enriched Si regions mentioned previously (Figure 24). The general corrosion on the free surface was similar to, but less intense, than that (described above) occurring under the crevices.

Material loss due to the impinging jet was by widely contrasting mechanisms. As can be seen in Figure 18, the effect of the impinging jet after one hour, was to cause a clear depression in the coating. Whereas the corrosion attack in static conditions resulted primarily in loss of matrix, the effect of the impinging jet was apparently associated with the hard phase. Particles of hard phase were clearly dislodged leaving voids in the matrix, shown in Figure 25 on the Co/50 coating. Around several of the hard phase particles there are clear indications of microcracking. It appears that, as may have been anticipated, mechanical material loss mechanisms contribute significantly under the liquid impinging jet whereas in corrosion attack, electrochemical material loss is via a contrasting preferential loss of the matrix material.

4 DISCUSSION

Characterisation of the spray-fused coatings confirmed that production via oxy-acetylene thermal spray and vacuum fusion processes results in a complex microstructure, the precise nature of which is not fully elucidated and does not appear to be reported in the thermal spray literature. This work has defined the nature of the strengthening phases and their distribution within the metallic matrix. In both the types of coatings considered here, the distribution of hard phase (comprising carbides, borides and nitrides) was relatively even throughout the thickness of the coating with the exception of a relatively thin region at the base of the coating,

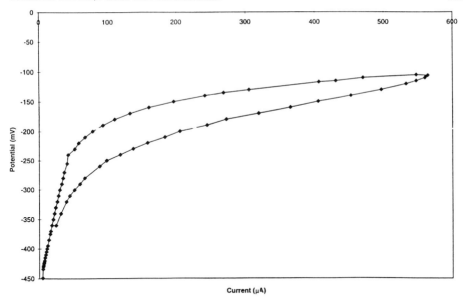

Figure 20 *High currents registered in the region positive to E_{corr} during anodic polarisation on Ni/40 under the impinging seawater jet*

adjacent to the substrate, which was free from precipitated hard particles. Extensive microanalysis of the coatings has confirmed that the principal strengthening phases in the Co-base coating are a mixed carbide, believed to be of the type M_6C as reported in the literature[20]. High concentration (around 55%) of W and significant levels of Co and of Cr were detected in this blocky hard phase. The vacuum anneal obviously results in significant interdiffusion both between the substrate and the coating, as indicated by the higher iron layer in the coating adjacent to the substrate/coating interface and by the presence of mixed carbides as hard phase.

The durability behaviour of thermal spray coatings in corrosive and erosive-corrosive environments is complex. In the severe conditions represented by seawater, the coatings investigated exhibit a degree of corrosion resistance and therefore provide a good measure of protection to a carbon steel substrate which, under the same conditions, undergoes active corrosion at significant rates. The corrosion behaviour of these cermet, spray-fused coatings displays many features comparable with traditional passive alloys such as stainless steels. Thus the coatings possess good corrosion resistance upon immediate exposure. However, continued immersion in seawater resulted, on all the coatings, in severe crevice corrosion in time periods similar to those that result in crevice corrosion of type 316L stainless steel. This study has not been of sufficient detail to allow estimates of the crevice corrosion propagation rates but the visual observations indicate that they are high. Thus, even the relatively large thickness (in the order of 1mm) of these coatings may not provide sufficient long-term protection. It is reasonable to suppose that the resistance to crevice corrosion will almost certainly be better in lower salinity waters but this requires verification.

In support of the visual observation detailed above, the electrochemical anodic polarisation scans, which on immediate exposure exhibited low currents in the potential region near to

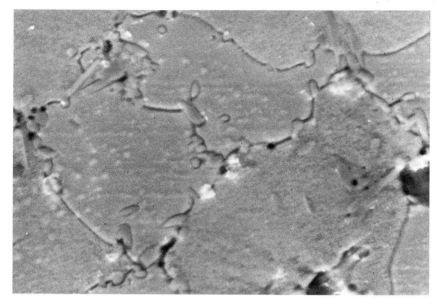

Figure 21 *SEM micrograph showing initiation of intergranular attack on Ni/50 after 64 days in seawater at the free corrosion potential*

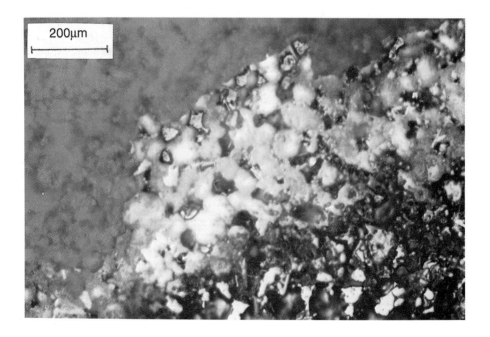

Figure 22 *Removal of entire grains from the region near to the lacomit/coating interface on Ni/50 after 40 days in seawater and anodic polarisation*

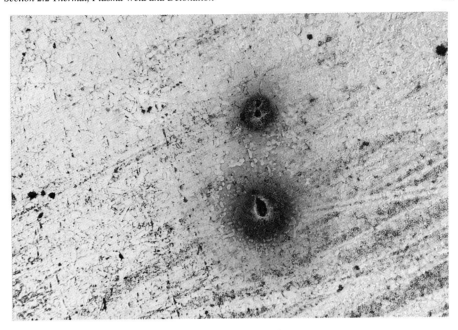

Figure 23 *Pitting attack on the free surface of Co/50 after 64 days in seawater at the free corrosion potential*

E_{corr}, after 40 days exposure, showed significantly larger currents due to localised attack. Another feature of the anodic polarisation, after longer-term immersion, was the higher maximum currents attained than after the 1 hour tests. This has been shown in previous work[18,19] to be associated with enhanced propagation and indeed the microscopy in this study has indicated the more severe crevice attack on the Ni-base coatings (which showed the highest maximum currents) compared to the Co-base coatings, after extended immersion.

An important feature is that the different coatings investigated displayed significant variations in corrosion behaviour both in the short- and the long-term. It is very likely that these variabilities are associated with the influence of microstructural components that comprise the coatings. Indeed, some evidence for corrosion concentrated at localised sites has been obtained in this study. Thus pitting attack was associated with the Si-rich areas in the Co-base materials. Also in the Ni-base coatings there was evidence of attack initiated at the grain boundaries, the severity of which was greater at the crevice regions at the coating/lacquer interface. This led to severe loss of matrix and resulting unsupported hard phase. The implications for retention of mechanical properties of these materials after periods in seawater are obvious.

It was also evident that, in the case of both the Co and the Ni coatings, the corrosion resistance of the higher hardness coating was somewhat lower than the material of reduced hardness. The exact reason for this trend might be associated with the concomitant increased porosity of the harder coating and increased hard phase concentration, at which localised attack was found on the Co coating. The resolution of the causes of this effect is important because of the possible implications of limitations in the achievement of both excellent corrosion and erosion resistance in one coating.

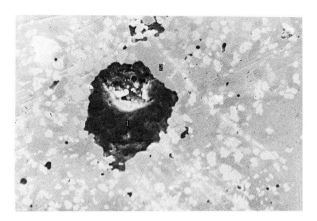

Figure 24 *Attack associated with the Si-rich region. Analysis at 1 – 29%Si, 0.5%Cr, 6%Fe, 0.25%Ni, 0.36%Co. Analysis at 2–9.5%Cr, 4.9%Fe, 23%Ni, 6.3%Si, 55%Co*

The preliminary study of erosion-corrosion behaviour in this work has also pointed to a complex phenomenon. The electrochemical measurements indicated a significant rôle of corrosion in the overall erosion-corrosion attack by providing anodic polarisation plots signifying active corrosion under the impinging jet. This was seen to be particularly so for the Co-base coatings. The microstructural evidence was that the erosion-corrosion mechanism involves microcracking at and around the large hard phases. It is of importance to determine in detail the relative behaviour of different hard phases in this respect and also the influence of more shallow impingement angles on the erosion-corrosion behaviour of these materials.

5 CONCLUSIONS

Preliminary results have been presented on the microstructural characteristics and durability performance of thermal-sprayed, vacuum annealed coatings. In terms of coating microstructure, this process results in a low porosity coating which, due to the metallurgical bond at the coating/surface interface, adheres well to the substrate. A number of different hard phases are incorporated in the metal-base matrix and interdiffusion between components in the coating ensures a dense compact layer.

Durability studies in seawater have revealed the following important aspects:

- the coatings studied exhibit good corrosion resistance upon immersion but, in a similar manner to passive alloys such as type 316L stainless steel, they suffer localised corrosion attack after moderate exposure periods,
- severe crevice attack occurred on all coatings after 40 days exposure and the Ni-base coatings appear to be less resistant to crevice corrosion initiation than cobalt-base coatings, and
- mechanisms of material loss are much different in static and impingement erosion conditions.

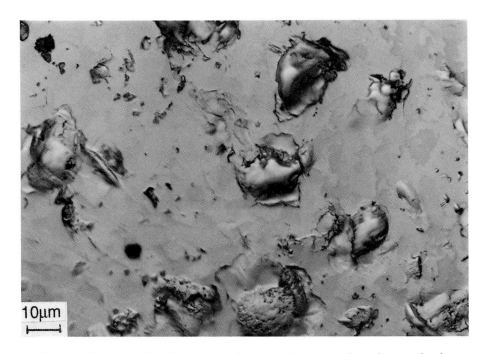

Figure 25 *Particles of hard phase dislodged from the matrix after 1 hour under the impinging seawater jet Co/50*

Evidence has been obtained of the association of localised attack to specific microstructural constituents such as Si-rich particles and matrix grain boundaries in static conditions and to the hard phase in erosion-corrosion conditions. Thus detailed studies refining such correlations would clearly contribute to optimisation of durability performance.

Acknowledgements

The authors acknowledge Professor B. F. Scott, Head of Mechanical Engineering, University of Glasgow and Professor J. E. L. Simmons, Head of Mechanical and Chemical Engineering, Heriot Watt University, Edinburgh for the provision of laboratory facilities.

References

1.   J. C. Cruzado, R. A. Hardin and H. McI. Clark, Proc. 7th Int. Conf on Erosion by Liquid and Solid Impact, 18-1, 1987.
2.   D. J. Greving, J. R. Shadley, Proc. Int. Thermal Spray Conference and Exposition, Orlando, Florida, 1992.
3.   E. J. Barette, 'High performance, severe service critical extraction ball valves, Proc. 8th National Thermal Spray Conference, 1995.
4.   K. Bremhorst and P. J. Flint, *Wear*, 1991, **145**, 123.

5. A. V. Levy, P. Crook, *Wear*, 1991, **151**, 337.
6. R. Martinella, G. De. Michele, V. Regis, Proc. 7th Int. Conf on Erosion by Liquid and Solid Impact, 18-1, 1987.
7. I. Iwasaki, S. C. Riemer, J. N. Orlich, *Wear*, 1985, **103**, 253.
8. D. H. Graham and A. Ball, *Wear*, 1989, **133**, 125.
9. G. R. Hoey J. S. Bednar, *Materials Performance*, April 1983
10. D. T. Gawne and I. R. Christie, *Metals and Materials*, Dec 1992, 646.
11. P. Siitonen, S. L. Chen, K. Niemi, P. Vuoristo, Proc. Int. Thermal Spray Conference and Exposition, Orlando, Florida, 1992.
12. K. Moriya, H. Tomino, Y. Kandaka, A. Ohmori, Proc. 7th National Thermal Spray Conference, 1994
13. Y. Kimura, T. Yoshioka, M. Kanazawa, Proc. 7th National Thermal Spray Conference, 1994.
14. P. Vuoristo, K. Niemi, A. Makela and T. Mantyla, Proc. 7th National Thermal Spray Conference, 1994.
15. R. Manuel, E. Yung, 'Metal seated ball valves with carbide coatings', Proc. 7th National Thermal Spray Conference, 1994.
16. A. A. Ashary and R. C. Tucker, 'Electrochemical and long-term corrosion studies of several alloys in bare condition and plasma sprayed with Cr_2O_3, *Surface and Coatings Technology*, 1990, **43/44**, p567–576.
17. A. A. Ashary and R. C. Tucker, CORROSION/93, Paper No. 24, 1993.
18. A. Neville and T. Hodgkiess, *Corrosion Science*, 1996, **38**, 6.
19. A. Neville, PhD Thesis, University of Glasgow, 1995.
20. A. Ashary and R. C. Tucker, *Surface and Coatings Technology*, 1991, **49**, 78.

2.2.5

Effects of Rare Earth Elements on the Plasma Nitriding of 38CrMoAl Steel

J. P. Shandong[1], H. Dong[2], T. Bell[2], F. Chen[3], Z. Mo[3], C. Wang[3] and Q. Pen[3]

[1]MECHANICAL AND ELECTRICAL EQUIPMENT CHIEF CORP., SHANDONG, CHINA

[2]UNIVERSITY OF BIRMINGHAM, BIRMINGHAM, UK

[3]SHANDONG UNIVERSITY OF TECHNOLOGY, SHANDONG, CHINA

1 INTRODUCTION

Notwithstanding the fact that new and innovative surface engineering technologies have been developed rapidly in recent years to meet ever increasing demands arising from some extreme applications, thermochemical treatments are by far the most widely used surface engineering techniques in industry today. This is largely because they are generally cheaper to operate, especially in high volume production, than most newly developed surface engineering technologies[1]. On the other hand, thermochemical treatments remain one of the main energy consumption sources in manufacturing industry. Therefore, how to speed up thermochemical processes presents a major challenge to surface engineers and researchers, which has served as a driving force for the development of plasma-assisted diffusion treatments.

More recently, there have been some attempts to speed up thermochemical processes, especially gaseous processes through incorporating rare-earth (RE) elements or compounds in these processes. Whilst the basic mechanisms involved remain unclear and thus controversial, general statements are made in the literature that the addition of RE to such thermochemical treatment processes as gas carburising[2], gas nitriding[3] and gas nitrocarburising[4] could effectively speed up these processes by 25–30% and improve, to some extent, the properties of the treated layers.

Surprisingly however, little attention has been paid to the study of the effect of RE on the plasma-assisted thermochemical process. Indeed, the addition of RE to plasma-assisted thermochemical processes could increase its scope of application. Firstly, the catalytic effect of RE may be more pronounced in the plasma assisted processes than in gaseous processes owing to the cathodic sputtering and hence high density of lattice defects in the surface layer, which are expected to promote the incorporation of RE in the surface layer. Secondly, due to their advantages of reduced processing time, low temperature processing and environmentally friendly nature, plasma-assisted thermochemical treatments are now widely used technically and in many situations, as economically viable alternatives to gaseous processes by a variety of manufacturing sectors[5].

Clearly, it is of great interest, both from a technical and scientific viewpoint, to investigate the effect of RE on the plasma-assisted thermochemical processes. The present work was thus undertaken to study the effect of RE on the plasma nitriding process of 38CrMoAl steel, and on the structure and abrasive wear resistance of the resultant layers, and finally to explore the catalytic mechanism involved.

2 EXPERIMENTAL DETAILS

2.1 Material

The substrate material used for plasma nitriding in the present investigation was a typical nitriding steel 38CrMoAl with the following chemical composition (wt.%): 0.37 Cr, 0.2 Mo, 0.84 Al, 0.4 Mn, 0.2 Si and balanced Fe. The steel was hardened (oil quenched from 940°C) and tempered (550°C for 1.5 h) to produce the desired core mechanical properties because plasma nitriding is usually used as the final procedure in manufacturing engineering components. Accordingly, a stable tempered structure, with a hardness of 350HV10 was obtained. Prior to plasma nitriding the specimens were ground to a surface finish of 0.7 μm (R_a).

2.2 Plasma Nitriding

The equipment employed in the work was a 25 kW commercial nitriding unit of model LD-25m which has been modified to facilitate the introduction of rare earth elements, La and Ce, to the plasma nitriding process. A bar made of rare earth compounds of La and Ce was hung in the chamber as an auxiliary cathode near the specimens, as schematically shown in Figure 1. Plasma nitriding was carried out in a cracked ammonia (NH_3) atmosphere at a temperature of 520°C. The treatment time for conventional plasma nitriding and plasma nitriding with RE (designated as *RE plasma nitriding*) was 8 h and 5 h respectively, in consideration of the catalytic effect of RE in thermomechanical treatment processes, which could reduce the treatment time by 25–30% as previously mentioned[2-4].

2.3 Characterisation of Nitrided Structure

The microstructure and metallurgical characteristics were examined and measured using a SEM and optical microscopy. To identify the phases present in the treated surface layer, standard X-ray diffraction analysis was carried out employing an X-ray diffractometer, with

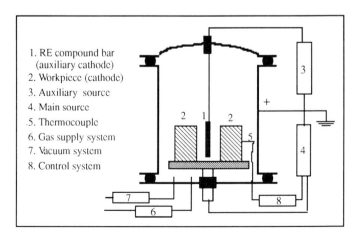

Figure 1 *Schematic diagram of plasma nitriding unit with RE auxiliary cathode*

copper radiation. The rare earth content and distribution in the RE plasma nitrided layer were measured using and EDX facility in the SEM. A microhardness tester, using a Vickers indenter, was employed for microhardness measurements. Hardness profiles of treated specimens were obtained on metallographic sections normal to the surface using a 50 gf load.

2.4 Abrasive Wear Testing

Abrasive wear resistance of conventionally nitrided and RE plasma nitrided 38CrMoAl was evaluated using an Amsler-type wear tester of model MM-200, with a wheel-on-wheel configuration. The rotation speed of the upper specimen wheel and the lower counterpart wheel were 360 and 400 rpm, respectively. Both wheels had the same geometry and size (40 mm outside diameter, 16 mm inside diameter and 10 mm thick) such that 10% sliding occurred between the two contact surfaces. The counterpart wheels made of AISI C1045 steel were boronised to produce a hard boride layer of 1500 $HV_{0.05}$. All tests were conducted at room temperature, with ambient humidity under various applied normal loads ranging from 300 to 1100 N for a rolling distance of 500 m. During the tests, abrasive particles comprising of iron borides of the size 10 to 50 μm and the hardness of 750 to 850 $HV_{0.05}$ were continuously introduced from a funnel down to the contact areas, as schematically shown in Figure 2.

3 RESULTS AND ANALYSIS

3.1 Metallurgical Characteristics

Metallographic examination revealed that both the RE plasma nitridied case and the conventionally plasma nitrided case have similar structural features, as with most thermochemically treated cases, comprising a thin compound layer adjacent to a diffusion zone. It is found that there is no appreciable difference in the morphology of the diffusion zone produced by conventional or RE plasma nitriding. But both the compound layer and the diffusion zone of the RE plasma nitrided case are much thicker than those of conventionally plasma nitrided case. The compound layers developed by the conventional and RE nitriding

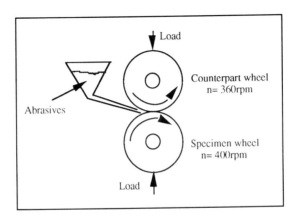

Figure 2 *Schematic diagram of abrasive wear tester*

processes were identified by XRD and the corresponding XRD charts are shown in Figure 3. It can be seen that the compound layer in the conventionally plasma nitrided case comprises both Fe_4N and Fe_3N, whilst that of the RE plasma nitrided specimen consists only of Fe_4N. It has been long recognised that a mono-phase (Fe_4N or Fe_3N) compound layer exhibits much better mechanical properties than a poly-base (Fe_4N or Fe_3N) compound layer. This is mainly because there are inherent stresses in the transition region between different structures in the poly-phase (Fe_4N or Fe_3N) compound layer, which may give rise to micro-cracks even if a slight external stress is applied[6]. Consequently, such a poly-phase compound is brittle and friable and it is prone to spall off[7]. Clearly, the addition of RE to the plasma nitriding process is expected to improve the mechanical properties of the compound layer.

Futhermore, EDX analysis results showed that RE elements mainly exist in the plasma nitrided compound layer, but there was no measurable amount of RE in the diffusion zone. The concentration of RE in the compound layer is represented in Table 1. It should be pointed out that the values listed in Table 1 can only be taken as a semi-quantitative result, since EDX is not suitable for the measurement of such light elements as nitrogen and oxygen. Even so,

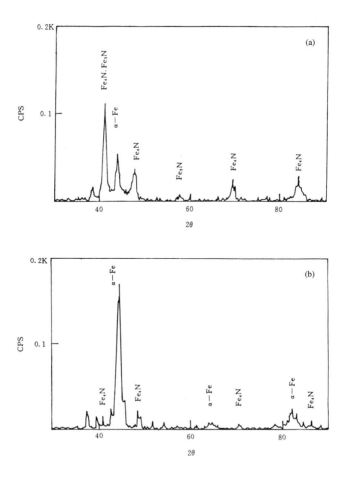

Figure 3 *XRD charts of (a) conventially nitrided and (b) RE plasma nitrided 38CrMoAl*

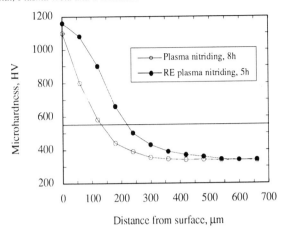

Figure 4 *Microhardness depth profiles for the conventionally plasma nitrided and RE plasma nitrided 38CrMoAl steel*

the result is still quite reliable because no RE elements were found in the conventionally plasma nitrided specimen.

3.2 Microhardness Distribution

The microhardness curves plotted against the distance from the surface for the conventionally plasma nitrided and RE plasma nitrided 38CrMoAl sections are shown in Figure 4. In the case of RE plasma nitriding, there is a peak hardness of 1170 $HV_{0.05}$ at the surface and the hardness decreases slowly with distance from the surface. The effective case depth, defined by the distance from the surface to the point where the hardness is 550 $HV_{0.05}$ was 224 µm. The total case depth at which the hardness reaches the value of the core hardness (350 $HV_{0.05}$) was 550 µm. The peak hardness of the conventional plasma nitrided specimen appears, as with the RE nitride specimen, at the surface with a hardness of 1100 $HV_{0.05}$. By contrast, the hardness of the conventionally plasma nitrided material decreases much more rapidly with distance from the surface, with the effective case depth and the total case depth being 134 and 350 µm respectively. Obviously, the addition of RE can significantly speed up the plasma nitriding process.

It is worth emphasising that the above case depth values correspond to the different treatment times, 8 h for conventional plasma nitriding process but only 5 h for the RE plasma nitriding process. Therefore, for the purpose of comparison, the extrapolated case depth values for the same treatment time were also estimated based on the assumption that parabolic dependence on time is applicable for both the plasma nitriding processes. It revealed that the effective case depth and total case depth will be 290 and 700 µm respectively when 38CrMoAl steel is RE

Table 1 *EDX analysis result of the element concentration in RE plasma nitrided 38CrMoAl steel*

Element	Ce	La	Cr	Mo	Al	Fe
Wt.%	1.18	0.76	2.39	0.73	1.86	92.38

Table 2 *Metallurgical characteristics of conventionally and RE plasma nitrided 38CrMoAl steel*

Process	Treatment conditions	Compound layer	Effective case depth	Total case depth
Plasma nitriding	520°C/8 h	1.5 μm	132 μm	350 μm
RE plasma nitriding	520°C/8 h	3.0 μm	224/290[*] μm	550/700[*] μm

[*]Extrapolated values for RE plasma nitriding at 520°C for 8 h.

plasma nitrided for 8 h at 520°C. Based on the above analyses, the metallurgical characteristics of both conventionally and RE plasma nitrided 38CrMoAl are summarised in Table 2.

The variation of weight loss in abrasive wear tests with the applied loads for conventionally plasma nitrided and RE plasma nitrided 38CrMoAl steel is shown in Figure 5. It can be seen that the weight loss, in accordance with abrasive wear law[8], increases with increase in applied loads for both the conventionally plasma nitrided and RE plasma nitrided 38CrMoAl steel. Clearly, the RE plasma nitrided steel exhibits better abrasive wear resistance over the conventionally plasma nitrided steel by a factor of 15–18%.

The morphology of the worn surfaces are shown in Figure 6. The worn surface of the conventionally plasma nitrided specimen was very rough, with deep and broad grooves as well as spalling features. By contrast, the worn surface of the RE plasma nitrided specimen was relatively smooth with some narrow grooves, which is in good agreement with weight loss results (Figure 5). This improved abrasive resistance can be mainly attributed to the higher hardness and deeper case depth, especially the deeper effective case depth, because the abrasive wear resistance is closely related to the hardness ratio between the abrasive and surface[8]. In addition, higher toughness of the mono-phase (Fe_4N) compound layer formed during the RE plasma nitriding process is believed to also contribute to the higher abrasive resistance since abrasive wear resistance is also associated with the toughness of the material[8].

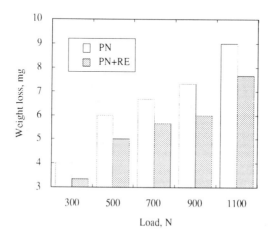

Figure 5 *Variations of abrasive wear as a function of applied loads for the conventionally plasma nitrided (PN) and RE plasma nitrided (PN+RE) 38CrMoAl steel*

4 DISCUSSION

It has been shown in the present work that the incorporation of RE elements, La and Ce, in the plasma nitriding of 38CrMoAl steel can significantly speed up the process, demonstrating the catalytic effect of these RE elements in the plasma thermochemical treatment. Qualitatively, the observed effects are in line with most other work on gaseous thermochemical treatments. However, the catalytic effect of RE is, quantitatively, much more pronounced in the plasma process than in the gaseous process. For example, Chen et al[3] found that addition of rare earth La to the gas nitriding process of 38CrMoAl steel could increase the case depth by up to 25% whilst in the present work an increase in the case depth as high as 100% has been observed when the same material was plasma nitrided with the addition of similar RE elements, implying that the plasma process could enhance the catalytic effect of RE elements. This could be explained by the different catalytic mechanisms occurring in the gaseous and plasma processes, which is believed to be closely related to the different nitriding mechanisms involved in these two processes. Although the plasma nitriding mechanism is a subject which is still open to debate, sputtering, ionisation, dissociation, ion bombardment and formation of vacancies and vacancy-ion pairs are undoubtedly the important reactions occurring in nearly all plasma processes[6,7].

(a)

(b)

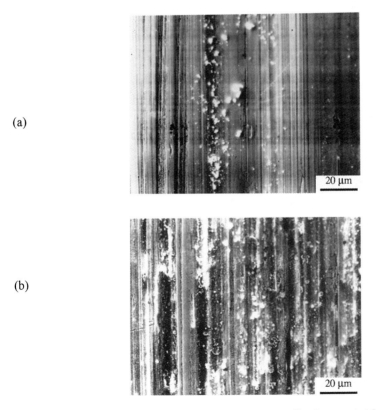

Figure 6 *Morpholgy of worn surfaces (a) conventionally plasma nitrided specimens and (b) RE plasma nitrided specimens*

Due to their special atomic shell structure and thus characteristic electronic configuration, most rare earth elements have high chemical activity and low electronegativity, and hence exhibit a very strong ionisation tendency. When sputtered by energetic nitrogen ions in the plasma, the surface of the RE compound bar releases RE-carrying species, most probably ions (Figure 7(a)). The activated RE ions detached from the RE compound bar are expected to increase the dissociation rate of NH_3 since these activated RE ions have extremely strong affinity with hydrogen[9]. Although some investigators[3] also claimed that RE could also increase the dissociation of NH_3 in gaseous processes, the phenomenon should be more pronounced in the plasma process than in the gaseous process owing to the higher activity of the RE-carrying species and the collision reaction in the plasma.

Furthermore, it is well known that the radii of most RE atoms are much bigger than those of iron, and consequently it is very difficult for RE atoms to diffuse in iron and steel from the viewpoint of classic diffusion theory. By contrast, this difficulty is expected to be partially overcome in plasma nitriding. Firstly, because of ion bombardment of nitrogen ions, radiation damage is caused in the surface layer (so-called *Winger effect*), thus forming vacancies, Frenkel pairs and dislocations[10]. According to the RE-vacancy model[11], the elastic strain energy caused by a RE atom dissolved singly in iron will be reduced by 50% if the RE atom is conjugated to a vacancy forming a RE-vacancy group. These vacancy-ion pairs are expected to migrate into the bulk by vacancy-substitutional diffusion at a much faster rate than the substitutional diffusion, especially for such large size atoms as RE atoms. Similarly, incorporation of RE ions in dislocations as well as grain boundaries will significantly reduce the elastic strain energy. Accordingly, ion bombardment-induced lattice defects favour the residence of RE in the surface. Secondly, RE ions may bombard the surface of the specimen being treated under the electrostatic force, which is thought to be beneficial for RE species to enter in the surface of the specimen (Figure 7(b)). In consideration of these two effects together, it is expected that RE species could be incorporated in the specimen surface more easily in a plasma process than in a gaseous process. But the depth remains quite limited since the diffusion rate of RE, though it is much faster than in gaseous processes, is still relatively low in view of the large size of the RE atoms. This hypothesis has been strongly supported by the fact that a high RE content (1.18% Ce and 0.75% La) appeared in the RE plasma nitrided layer but it was distributed only in the first several microns.

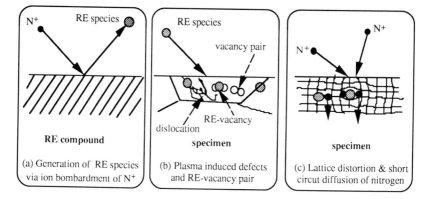

Figure 7 *Plasma enhanced RE catalytic mechanism involved in the RE plasma nitriding process*

Once the RE species enter the specimen surface, the surrounding lattices will be heavily distorted due to the great difference in size between the RE species and the iron atoms, which is beneficial for the diffusion of the alloying species (nitrogen in the present work) further into the core, by way of short circuit diffusion[12], as schematically shown in Figure 7 (c). As a result, a high nitrogen concentration forms on the surface in very short time, providing an extremely high nitrogen potential and a steep nitrogen gradient for the rapid diffusion of nitrogen into the material. Again, this speculation is positively supported by the experimental result that Fe_3N was not formed in the compound layer in the RE plasma nitrided material owing to the fast short-circuit diffusion of nitrogen inwards to the core.

5 CONCLUSIONS

1. Introduction of rare earths into the plasma nitriding process of 38CrMoAl steel can be successfully performed by directly hanging a rare earth compound bar as an auxiliary cathode in the treatment furnace near the specimens. Rare earth-carrying species, more likely ions, can diffuse into the nitriding case and mainly exist in the compound layer.
2. Rare earth elements La and Ce have a significant catalytic effect on the plasma nitriding process for 38CrMoAl steel, and can increase the case depth by more than 100% as compared with conventional plasma nitriding for the same treatment time.
3. The addition of RE elements to the plasma nitriding process promotes the formation of the mono-phase Fe_4N compound layer, which possesses desired mechanical properties in terms of high hardness and adequate toughness.
4. The RE plasma nitriding of 38CrMoAl steel can increase the abrasive wear resistance by 15–18% over the conventional plasma nitrided specimen. The worn surface of conventional plasma nitrided specimens exhibits deep abrasive grooves as well as brittle spalling, whilst that of the RE plasma nitrided specimens shows mild abrasive wear features.

Acknowledgements

The authors wish to thank their colleagues in the Wolfson Institute for Surface Engineering at Birmingham University – particularly Dr. Y. Sun – for invaluable discussions on many aspects of the work.

References

1. Department of Trade and Industry, 'Wear Resistant Surfaces in Engineering', Her Majesty's Stationary Office, London, 1986, p. 55.
2. S. Wang, L. Wang and A. Yang, *Heat Treat. Met.* (China), 1988, **3**,52 (in Chinese).
3. X. Chen, D. Ding, M. Lu, J. Liu and N. Chen, *Heat Treat. Met.* (China), 1990, **12**, 3 (in Chinese).
4. X. Chen, D. Ding, M. Lu, J. Liu and N. Chen, *J. Rare Earth*, 1991, **9**, 129.
5. In Ref. 1, p. 63.
6. B. Edenhofer, *Heat Treatment of Metals*, 1974, **2**, 59.
7. T. Bell, T. Rees and V. Korotchenko, Proc. Conf. on Ion Plating and Allied Tech., June 1977, p. 230.

8. I. M. Hutchings, 'Tribology – Friction and Wear of Engineering Materials', Edward Arnold, London, 1992.
9. K. A. Gschneider and L. Eyring, 'The Handbook on the Physics and Chemistry of Rare Earths', Vol. 12, North Holland, Amsterdam, 1989.
10. N. L. Peterson and S. D. Harkness, 'Radiation Damage in Metals' ASM, Metals Park, Ohio, 1976.
11. Z. Liu and Z. Lu, in G. Xi, J. Xiao, Z. Yu and M. Chen (eds), Proc. 2nd Int. Conf. on RE Development and Application, International Academic Publisher, Beijing, China, 1991, p. 925.
12. R. E. Reed-Hill, 'Physical Metallurgy Principles' (2nd edition) D. Van Nostrand Company, London, 1973.

2.2.6
APNEP – A New Form Of Non-Equilibrium Plasma, Operating At Atmospheric Pressure

N. P. Wright

E. A. TECHNOLOGY, CAPENHURST, CHESHIRE, UK

1 INTRODUCTION

The activating properties of non-equilibrium plasmas are increasingly exploited in many industrial sectors, including engineering, optics, plastics and electronics. Applications include: surface modification of plastics in packaging, preparation of plastic components for painting, anti-corrosion and wear resistant treatments for engineering components, etching and deposition in the semiconductor industry, component cleaning, effluent treatment, and many others.

Non-equilibrium plasmas are ionised gases of low thermal energy, which also contain far more energetic electrons. This unique method of energy delivery allows them to promote entirely new processes and enhance existing ones, generating advanced surface properties and reducing reaction temperatures. They also have environmental advantages: through providing solvent replacement, energy efficiency, and effluent abatement. They are the subject of much research worldwide, and substantial further growth is anticipated[1].

Although they can bring great benefits to industrial processes, non-equilibrium plasmas are normally produced in vacuum, incurring the penalties of high capital cost and inefficient batch processing. These drawbacks are overcome by the EA Technology Atmospheric Pressure Non-Equilibrium Plasma (APNEP), which is the subject of a patent application. Its main advantages are:
- operation at atmospheric pressure, giving continuous processing and high throughput, and
- utilisation of simple, inexpensive components.

These give APNEP the potential to improve the prospects of applications currently using vacuum plasmas, and to bring the benefits of non-equilibrium plasma processing to new areas where vacuum methods are not currently viable.

2 PLASMA PROCESSES

Plasmas contain various combinations of ions, electrons and neutral species. Excitation is by DC, radio frequency or microwave frequency power sources. Conventionally, they are classified into one of two types : thermal or non-thermal (also known as "cool", "cold", or "non-equilibrium").

Electrical breakdown at pressures higher than rough vacuum requires an intense field. Combined with the high density, this causes substantial energy exchange between particles, producing extensive ionization and high temperatures. Consequently, the resultant species are in local thermal equilibrium. These "thermal plasmas" are used as sources of high power and temperature, usually at atmospheric pressure, in processes such as metal cutting, welding, thermal spraying of ceramics, melting of metals and glass, and waste destruction.

At low pressure, less intense fields are required. The reduced frequency of collisions between particles gives lower temperatures and allows acceleration of the electrons, because of their greater mobility, to higher kinetic energies than if they were in equilibrium with the bulk of the plasma. Their collisions further disturb the thermal equilibrium, giving a "non-thermal" (non-equilibrium) plasma. These energetic electrons and their collision products activate unique reactions which thermal plasmas are incapable of promoting. As part of a comparatively cool bulk medium, they can lower reaction temperatures and facilitate processing of heat sensitive materials[2,3]. Interactions with the substrate as well as the gas phase reactants, enable non-equilibrium plasmas to generate advanced surface properties unachievable by other means.

APNEP has useful attributes of both types of plasma, but without the drawbacks. It uniquely combines the advantages of the atmospheric pressure operation of thermal plasmas with the activating properties of non-equilibrium plasmas, which are crucial to many surface engineering and effluent treatment applications. Examples of these uses include: semiconductor etching and deposition[4]; hard finishes for cutting tools by carburising and nitriding[5]; deposition of ceramic coatings for hard, wear resistant and protective tribological surfaces[6]; optical coatings on glass; plasma polymerisation to give high performance surfaces on cheap substrates[7]; surface activation to assist printing and painting; surface modification to enhance hydrophobicity; plating of plastics. Evidence to date suggests that APNEP has the necessary properties to address all of these areas.

3 DESCRIPTION OF APNEP

The basic features of APNEP are:
- a multi-mode microwave cavity adapted from a commercial oven;
- a power source adapted from commercial microwave components;
- specially designed plasma containment;
- plasma initiation equipment.

Electrons and ions generated by initiation equipment absorb the microwave energy and rapidly develop into a diffuse glowing plasma, which is sustained on cessation of the initiation process. The glow indicates continual recombination and relaxation of excited species, which are simultaneously replenished by the microwave field. Modifications to the proprietary microwave oven are not major, and the widespread availability of industrial microwave components means that the device will retain its low cost even after development for industrial use.

This plasma is unusual in that it does not appear to fall wholly into either of the categories of plasma consistent with conventional experience. It is sustained continuously at atmospheric pressure, typical circumstances for generation of a thermal plasma, yet it operates at much lower temperatures and has the appearance of a glow discharge normally associated with non-thermal plasmas. These factors suggest that it has the activating properties required for non-equilibrium surface processing, while its atmospheric operation renders it more amenable to continuous processing than competing vacuum plasmas. This combination, allied to the cheap construction, gives this plasma great promise in surface engineering.

In normal atmosphere the plasma glows quite strongly in the visible region, but appears to be separated from its containment vessel by a less bright sheath region of several millimetres thickness. This is a characteristic of glow discharges in which a substrate or wall in contact with the plasma tends to be at a lower potential than the plasma itself. Positive ions move preferentially towards this lower potential, while electrons are repelled, disturbing the macroscopic electrical neutrality in the space near the substrate. Only a few highly excited electrons will be able to penetrate this region, and since the plasma glow is caused by relaxation of atoms excited by electron impact, the comparative paucity of such events in the sheath region gives it its darker appearance.

Because APNEP still contains significant thermal energy, a wide range of processing temperatures is possible by applying fluid dynamics principles. Treatment can take place either within or outside the microwave cavity. In processing inside the cavity, the energy from the field is still available at the substrate surface. This gives better prospects for a high quality surface since deposition from the gas/plasma phase takes place as close as possible to the surface, although more sensitive materials may need better protection from the main discharge, while others may be incompatible with the microwave field. Processing outside the cavity, however, permits simpler engineering for presentation and manipulation of the sample, and measures can still be taken to maintain energisation of species within the treatment stream.

For low processing temperatures, gases are fed through the main ionisation region to generate a cooler "afterglow", transporting excited species to the substrate at a remote "downstream" position, near to which the reactants are injected. This has many advantages :

- heat sensitive materials and reactants can be used;
- plasma processes take place close to the surface giving superior properties;
- damage to the substrate from UV and ion bombardment is minimised;
- decoupling of surface processes from plasma production improves control; and
- reaction products do not accumulate in the plasma vessel where they could hinder performance.

Such principles are increasingly adopted in vacuum plasmas in order to minimise substrate damage[6,7,8]. Where higher inputs of thermal energy are tolerated, or even desirable, APNEP can be used directly, with materials and reactants adjacent to the plasma source.

Theory suggests that at high pressure the temperature should be too great, and the Debye length and electron mean free path too short, for useful processing except in the main discharge. That this is not so may be due to several factors, including : the presence of long-lived, excited, non-ionic species[9]; non-equilibrium conditions in the space charge "sheath" region near the flask surface[4]; and the manner in which the magnetrons are driven.

The unusual combination of non-equilibrium conditions and atmospheric operation, as offered by APNEP, is crucial in enabling industry to widely exploit the benefits of plasma processing[1]. Some specialised methods do exist, such as corona discharges for pretreatment of packaging film. Other techniques under research include pulsed RF plasmas, surface wave plasmas, annular slotted waveguide applicators, focussed microwave pulses and lasers in a resonant cavity. While all these approaches are of value, it is felt that APNEP, through its comparative simplicity and flexibility, offers excellent prospects for industrial implementation.

4 INITIAL THEORETICAL ASSESSMENT

The fundamental nature of the plasma, either thermal or non-thermal, was determined from thermal analysis based on simple observations. Because the microwave environment renders some measurements difficult, the initial theoretical assessment was based on estimated ranges of input data. No iterative procedures were involved. Instead, the sensitivity of the plasma model to these ranges was examined. From the resulting matrices of plasma parameters, the realistic areas of operation could be assessed with reference to experimental experience.

The calculation process consisted of two stages : estimation of the plasma temperature followed by calculation of plasma conditions based on this value. The dominant heat transfer mechanisms considered are applied within a framework of data estimated to represent ranges of realistic conditions. To facilitate an analytical approach, some simplifying assumptions are made: the plasma is uniform; it is surrounded by a sheath region which is in contact with the spherical flask; and the temperature is uniform across the outer surface of the plasma ball.

In this simplified model, incident energy absorbed by the plasma loses heat via the following route:

1. Conduction from the plasma surface through the plasma sheath,
2. Conduction through the flask wall, and
3. Convection from the flask surface.

As other loss mechanisms are considered negligible, the heat flow is constant through the three regions. It is convenient to treat the stages in the reverse of this order. The vessel surface temperature is estimated, and the corresponding convective heat flux evaluated to establish the heat flow through the system. The appropriate equation for the heat loss is :

$$Q = h_m \prod D^2 \left(T_o - T_\infty \right)$$

where Q = heat loss; D = diameter; T_o = surface temperature; T_∞ = "sink" temperature and h_m is the heat transfer coefficient, derived from the dimensionless quantities: Nusselt number, *Nu*, Grashof Number, *Gr* and Prandtl Number, *Pr*. These are related by the semi-empirical relationships:

$$Nu = h_m \frac{D}{k_t} = 2 + 0.59 \times Gr^{\frac{1}{4}} \times Pr^{\frac{1}{4}}$$

where k_t = thermal conductivity;

$$Gr = D^3 \rho^2 g \beta \frac{\left(T_0 - T_\infty \right)}{\mu^2}$$

where: r = density; g = acceleration due to gravity; b = coefficient of thermal expansion; and m = dynamic viscosity;

$$Pr = C_p \frac{\mu}{k_t} \quad Pr = Cp \ m/k_t$$

where: C_p = specific heat at constant pressure.

Using the outer surface temperature as a starting point, the heat flow is used to successively determine the temperatures at the flask/sheath interface and at the sheath/plasma interface, which in this model corresponds to the plasma temperature itself. The surface temperature of the flask depends on the rate of convective heat loss, which is equal to the absorbed microwave power. A range of surface temperatures was assumed in order to calculate a corresponding range of flows. The matrix of results generated could then be reduced to a much smaller set of realistic conditions by comparison with observations. There is little temperature drop across the flask wall because of its very low thermal resistance. For each combination of heat flow and flask temperature generated, the temperature at the perimeter of the plasma is calculated from conduction through the plasma sheath. Again, a range of sheath thicknesses was used in each case, in order to straddle the estimated value. In practice, the result is not too sensitive, changing by only 60% for a five-fold increase in thickness.

The fundamental requirement is to establish whether the plasma is thermal or non-thermal. The two types are described by different theories, so it is valid to apply either one, and then to test whether it is consistent with the observed phenomena. A thermal model was tested here since this description is more readily applied to the available data. The range of feasible plasma temperatures established earlier from the heat transfer model was used to calculate fundamental parameters which serve as criteria for assessing the viability of the plasma. If these suggest that the plasma can be sustained from thermal mechanisms alone then the description is valid: if not, then it must be concluded that non-thermal effects play a significant rôle.

It was calculated that between 30% and 50% of the incident microwave energy must be absorbed in the plasma to maintain flask surface temperatures in the range 700 to 1200°C as estimated. The highest corresponding plasma temperature was 2100°C, if assumed uniform. If uniformity of energy absorption is assumed instead, however, the temperature will rise towards the centre, reaching about 4000°C. A position between the two is most likely. Comparison of observed cooling rates with this theory at low temperatures suggests that the rate of heat loss is underestimated, leading to a slight exaggeration of plasma temperature.

In developing a simple plasma model, it was assumed that local thermal equilibrium is established, that the plasma is uniform, and that single charge ionization dominates. The thermally generated ion density, equal to the electron density for single ionization, was calculated from an appropriate version of the Saha equation[10,11] :

$$\frac{n_e^2}{n_n} = g \left\{ 2\pi m_e K \frac{T_e}{h^3} \right\}^{\frac{3}{2}} e^{-\frac{W_i}{KT_e}}$$

where: n_e = electron number density; n_n = neutral number density; T_e = electron "temperature"; W_i = ionisation energy; m_e = electron mass; h = Planck's constant; and K = Boltzmann's constant. Having established the electron density, the collision frequency and plasma frequency can be calculated. These are significant in establishing the transfer of energy from the field to the plasma via dielectric loss, and can be determined using semi-empirical relationships[11,12]:

plasma electron frequency, $\omega_{pe} = 8.98 \times 10^3 \sqrt{n_e}$ sec^{-1} ;

electron collision frequency, $\upsilon_e = 2.91 \times 10^{-6} n_e \ln(\Lambda) T_e^{-\frac{3}{2}}$ sec^{-1}

$\ln(\Lambda)$ is the Coulomb logarithm. It is derived from the impact parameter for Coulomb collisions and the Debye length, and is typically of the order of ten[12].

The permittivity of the plasma is then given by :

$$\varepsilon_p = \varepsilon'_p - i\varepsilon''_p = \frac{1 - \omega^2_{pe}}{\{\omega(\omega - i\upsilon_e)\}}$$

where ε_p = plasma permittivity, and ω = microwave frequency. From this, the imaginary part can be used to determine the relationship between power density and electric field :

$$P_v = \omega_p \varepsilon_0 \varepsilon''_p E^2$$

where: ε_0 = permittivity of free space; P_v = power density; E = electric field strength.

Ion density is a fundamental parameter, crucial to assessing the nature of the plasma. Even for the highest temperature the value obtained is only 6×10^8 cm^{-3}, which is much lower than expected for a thermal plasma. Support of this ion density at the observed decay rate can only dissipate a minute fraction of the power absorbed. There must therefore be substantial non thermal mechanisms involved.

For most of the range studied, the plasma frequency is below the microwave frequency. The field can therefore penetrate the plasma and energy transfer can take place. A reasonable estimate of the incident power density, however, results in a required field strength of about 5 $\times 10^{11}$ V/m, which is several orders of magnitude greater than possible in equipment of this nature.

Recombination rate and maximum ionisation rate can both be estimated from the plasma temperature and ion density. This ionisation rate is the maximum possible if all the electron kinetic energy were transferred for each collision, although in reality only a proportion will contribute. This maximum value is significantly higher than the recombination rate, which, in principle, means that the possibility of a thermal plasma cannot be discounted solely on these grounds. Individually, however, the rates are too low to be consistent with conventional experience of thermal plasmas.

Overall, the calculations indicate that thermal effects are insufficient to sustain the plasma. Because of this failure of the thermal plasma model to explain the observed phenomena, it is concluded that APNEP is essentially non-thermal, the electrons acquiring additional energy beyond equilibrium levels from the electromagnetic field. This suggests that the plasma is suitable for treating heat sensitive materials, as well as for a wide variety of processes involving the type of activated gas phase chemistry associated with non-thermal plasmas. This is supported by APNEP's ability in two applications where non-equilibrium properties are essential: coating of polymers and destruction of volatile organic compounds.

5 COATING OF POLYMERS

The non-equilibrium component of the plasma can promote gas phase reactions at reduced temperatures, allowing coating of heat sensitive materials. APNEP still has a significant thermal content, however, which requires careful management. Low temperature processing is achieved by using gas streams to carry activated species from the main discharge to a substrate located at a cooler position. Other configurations include larger area treatment by direct contact with cooler regions of the main discharge. These various arrangements cover substrate positions both within and outside the microwave cavity, each option requiring special consideration for the application in question. In coating trials, inorganic barrier layers of titanium dioxide have been applied to filled polymethyl methacrylate and epoxy insulating materials to improve their weather resistance. Two phases of the work have been completed.

In the first phase, disc samples of 8mm diameter were suspended in an exit tube from the main plasma region, within the microwave cavity. The sample position and reactant flow rates were adjusted to achieve good coating quality while preventing thermal damage. Encouraging results were obtained. Examination by optical microscopy revealed typical coating thicknesses of one micron. Applied in periods of up to 20 seconds, this represents a very high deposition rate in comparison with vacuum plasma methods. Coatings were reasonably uniform, with X-ray diffraction analysis confirming significant rutile content. The roughness of the original composite substrate was generally reduced by the coating : sometimes by as much as 50%. Surface energy, as measured by water droplet contact angle, was equal to or less than the virgin material. In some cases, contact angle increased from 70° to 120°.

In the second phase, the work was scaled-up to coat discs of 7cm diameter which, in contrast to the earlier work, were treated outside the cavity. An activated stream is carried from the plasma and allowed to emerge onto the sample surface. The aperture is smaller than the sample, which is manipulated over the application point to give uniform coverage. In conjunction with appropriate gas flows, this also gives good control over the substrate

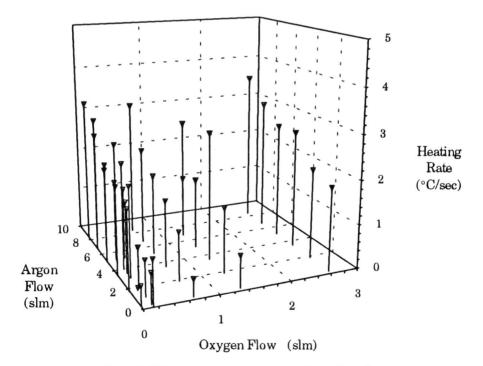

Figure 1 *Substrate heating rate as a function of gas flows*

temperature, which is well illustrated by the capability to limit temperature rises to less than 5°C/sec in this case (Figure 1).

Since deposition rates are high at atmospheric pressure, such heating rates permit ample coating time even for heat sensitive substrates.

To assess the effectiveness of the coating as a barrier layer, analysis and testing were repeated on samples aged by UV light in a moist environment to simulate weathering. The results in Tables 1 and 2 indicate few significant differences between coated and uncoated samples, although the coating does appear to give better retention of a low energy surface after ageing.

These larger samples were suitable for subjecting to functional tests relating to outdoor insulator applications. In the inclined plane tracking test (IEC test specification 587), samples are subjected to electrical stress across the surface while contaminated with a weak electrolyte. Three sets of conditions are imposed, for 6 hours each, at increasing levels of voltage and electrolyte flow rate, with a maximum allowable surface current specified. This is limited by the surface resistance of the wetted sample, and will increase if the surface is degraded. The results are summarised in Table 3. Before ageing, both uncoated and coated samples lasted the specified time at level one. At level two, neither managed to survive for an hour, but the coated samples showed some initial durability at level three. It is the comparison in the aged condition which is most relevant in assessing the effectiveness of the coating. Although neither type survived the necessary six hours at level one, the coated samples lasted 50% longer than the uncoated ones, and survived for half an hour at level two.

Organic effluent is increasingly a concern in a number of industries as a result of the Environmental Protection Act of 1990. Abatement equipment is expensive, and frequently beyond the means of the small operators who constitute a large proportion of such industries. Thermal plasmas are inefficient in destroying VOCs, particularly at the low concentrations commonly produced, while the high temperatures cause recombination. Non-equilibrium plasmas contain the necessary species, but need vacuum operation, which is not compatible with the high flow continuous VOC streams typically requiring abatement. APNEP is well-suited to this application, however, due to its atmospheric pressure operation, high throughput, and low cost.

Initial trials have demonstrated its potential. In separate tests, acetone, toluene and trichloroethylene, at concentrations from 200 to 1000vpm, were destroyed with greater than 97% efficiency. Since these early experiments, the maximum flow rate has been rapidly increased from 14 to 120 standard litres per minute, the only limitation being the capacity of the flow meter. This early success, achieved with the help of fluid dynamics modelling, indicates that

Table 1 *Surface roughness measurements*

Surface	Roughness – individual measurements (μm)
Epoxy	0.67, 1.0, 0.47, 0.70
Epoxy after ageing	0.50, 0.47, 0.8, 1.1
Coated epoxy	0.77, 0.35, 0.45, 0.25
Coated epoxy after ageing	0.47, 0.42, 1.30, 1.05

Table 2 *Surface energy comparison by water droplet contact angle*

Uncoated		Uncoated and aged		Coated		Coated and aged	
Initial	After 1 minute	Initial	After 1 minute	Initial	After 1 minute	Initial	After 1 minute
117	121	118	127	103	117	95	117
123	127	91	112	162	161	126	141
115	123	105	123	126	127	122	132
Mean 118	Mean 124	Mean 105	Mean 114	Mean 130	Mean 135	Mean 114	Mean 130

further order of magnitude increases are readily achievable. The inherent low cost of the equipment also allows multiple arrangements in order to further increase flow rates. Parallel investigations into improving efficiency through microwave system design are being carried out. Areas of study include waveguide structures, resonant cavities and metal plasma confinement structures.

6 FUNDAMENTAL STUDY OF THE PLASMA PHYSICS OF APNEP

The development of APNEP is underpinned by a fundamental investigation which is part of the Post Graduate Training Partnership run by EA Technology and UMIST with support from the DTI. Conclusions so far are:
- The power primarily influences the volume of the plasma.
- The rate of decay of optical emissions depends on pressure.
- The main energy losses from the system are by diffusion to the walls and recombination of the charged species.
- The plasma can extend more than ten centimetres beyond the microwave cavity.
- Microwaves are absorbed mainly in the surface of the plasma because of its high density.
- The plasma is predominantly non-equilibrium.

Table 3 *Results of the surface tracking tests*

Insulator finish	Time for which leakage current remained within specification		
	At level 1	At level 2	At level 3
Uncoated	>6hrs	40mins	-
Coated	>6hrs	20mins	15mins
Uncoated and aged	1hr	-	-
Coated and aged	1hr 20mins	30mins	-

Future work will identify the main species and determine more precisely the ion density and electron energy.

7 CONCLUSIONS

APNEP is at an early stage of development, but will progress rapidly towards industrial applications. In developing the engineering for efficiency, reliability, flexibility and control, the following issues will be investigated:

- Optimization of cavity size and shape.
- Modifications to the microwave source to maintain stability at all power levels.
- Increasing the non-equilibrium proportion of the plasma – for example by vessel design and flow control, and further electrical modifications.
- Determination of the relationships between flow rate, power and plasma shape.
- Investigation of downstream processing.
- Comparison of processing within and outside the microwave cavity – to optimise the balance, for specific applications, between ease of engineering, and retention of energy in the plasma stream.
- Large area application. Certain flow patterns have been devised which can allow large diameter plasma exit conduits, giving much greater treatment areas.
- Development of the plasma initiator system – for control and flexibility.
- Treatment of three-dimensional shapes – by vessel design, applicator design, and flow control.
- Durability engineering of the system.
- Development of relevant monitoring techniques. These may include spectroscopic methods, or inferred parameters from simpler measurements as appropriate.

The preliminary results indicate the potential of APNEP in one particular application. This experience suggests that it is equally applicable to other coating and substrate systems. The initial success augurs well for its implementation in a wide variety of surface engineering applications currently employing vacuum plasma methods.

Acknowledgements

The author wishes to acknowledge the contribution of his colleague, Dr Xiaoming Duan, who discovered this form of plasma. The work of Mr H Potts, Prof M Rusbridge and Prof J Hugill of UMIST for the fundamental investigations is also gratefully acknowledged.

References

1. S. P. Howlett, S. P. Timothy and D. A. J. Vaughan, 'Industrial Plasmas: Focusing Skills On Global Opportunities', CEST, 1992.
2. Hollahan, J.R. and Rosler, R.S. in 'Thin Film Processes' (L. J. Vossen, and W. Kern, eds) Ch IV-1, Academic Press, 1978.
3. O. Knotec, F. Löffler, and Kramer, G. Third International Conference On Plasma

Surface Engineering, Garmisch-Partenkirchen, Germany, October 26-29, 1992. *Surface And Coatings Technology*, **59**, 14.

4. Gerretsen, A., Lottermoser, L., Müller, J. Third International Conference On Plasma Surface Engineering, Garmisch-Partenkirchen, Germany, October 26–29, 1992. *Surface And Coatings Technology*, **59**, 212.

5. Hombeck, F., Oppel, W., Rembges, W. and Leverkusen. Proceedings 1st International Conference on Plasma Surface Engineering, DGM, 1988.

6. J. Hamerisch, G. Kirchner, T. Kelly, V. Mernagh, R. Koekoek and L. McDonnel, Third International Conference On Plasma Surface Engineering, Garmisch-Partenkirchen, Germany, October 26–29, 1992. *Surface And Coatings Technology*, **60**, 566.

7. B. Robinson, P. D. Hoh, P. Madakson, T. N. Nguyen, and S. A. Shivashankar, Plasma Processing And Synthesis Of Materials Symposium, Anaheim, California, USA, 21–23 April, 1987, MRS Proceedings Vol. 98, p. 313.

8. W. Kulisch, M. Witt, H. J. Frenck and R. Kassing, *Materials Science And Engineering*, 1991, **A140**, 715.

9. W. Kulisch,. Third International Conference On Plasma Surface Engineering, Garmisch-Partenkirchen, Germany, October 26-29, 1992. *Surface And Coatings Technology*, **59**, 193.

10. E. Nasser, 'Fundamentals Of Gaseous Ionisation And Plasma Electronics', Wiley, 1971.

11. Anders, A. 'A Formulary For Plasma Physics', Akademie-Verlag Berlin, 1990.

12. Anderson, H.L. "A Physicist's Desk Reference", American Institute of Physics, 1989.

2.2.7

Scanning Reference Electrode Study of the Corrosion of Plasma Sprayed Stainless Steel Coatings

D. T. Gawne, Z. Dou and Y. Bao

SCHOOL OF ENGINEERING SYSTEMS AND DESIGN, SOUTH BANK UNIVERSITY, LONDON, UK

1 INTRODUCTION

Thermally sprayed stainless steels coatings are widely used to confer corrosion resistance to steel components. However, practical experience indicates that some corrosion still occurs under severe environmental conditions. This paper is directed at investigating the corrosion mechanism of the coatings with the aim of improving their protective ability.

Conventional techniques for the study of corrosion usually require the periodic removal of the metal from the corroding environment. Research on the repassivation of stainless steels[1] has shown that the removal or replacement of these materials in the corroding medium may exert a considerable effect upon the electrochemical activity. In particular, existing pits may repassivate and new pits may initiate. The scanning reference electrode technique (SRET), which is an in-situ method of monitoring electrochemical activity[2,3], immediately overcomes this limitation.

2 EXPERIMENTAL DETAILS

The feedstock powder used was stainless steel (41C grade) supplied by Sulzer-Metco Ltd. The particle size distribution of the powder is: <45µm: 3wt.%; 45–53 µm: 6wt.%; 53–106 µm: 71wt.% and >106 µm: 20wt.% and an SEM micrograph of the powder is given in Figure 1. The substrate used was a plain carbon steel (EN8) plate of thickness 6 mm, which was cut to 25×30mm, degreased and grit blasted with alumina grit (Metcolite C, Sulzer–Metco Ltd) under a blast pressure of 4bar and a blast distance of 150 mm to give a surface roughness of 7 µm Ra. Plasma spraying was undertaken using a Metco plasma spray system with an MBN torch, MCN control unit, 4MP powder feed unit and fluidized hopper.

The corrosion investigation was carried by using an electrochemical impedance analyzer (Model 6310, EG&G (UK) Ltd.), which includes a computer to perform the control function and an electrochemical cell. A three-electrode electrochemical cell was used in the investigation, in which the sample was used as the working electrode, with saturated calomel and platinum electrodes as reference and counter electrodes. All potentials were measured with respect to a saturated calomel electrode(SCE). The corrosion properties were studied in three solutions: 3.5 wt% NaCl, 2×10^{-3}M HCl and tap water. The test temperature was controlled at 25 °C. Each sample was degreased with acetone and coated with lacquer to give an exposed area of

1cm² before immersing into solution. Both the potentiodynamic test and the Tafel test were carried out following the ASTM G5-82 procedure[4], using a potentiodynamic potential sweep rate of 0.6 V/h (±5%) recording the current continuously with the change in potential. The general characteristics were obtained from the potentiodynamic tests and the corrosion rates from Tafel tests.

The scanning reference electrode technique (Model SP 100 EG&G Ltd.) was used to study the localised corrosion behaviour of the coatings. The corrosion current-time relationships of the coatings were obtained from the chronoamperometry test (CA), which controls the applied potential at a constant value of –200mV and measures the current with respect of time.

3 RESULTS AND DISCUSSION

Figure 2 shows the polarization plots for the steel substrate, bulk stainless steel (of the same composition as the feedstock powder) and the coatings in a 3.5wt% NaCl solution. It shows that stainless steel is, as expected , more noble than the steel substrate ($E_{corr,ss} > E_{corr,Fe}$) and thus acts as a cathodic coating. The polarization curve for the coatings lie in between stainless steel and the steel substrate, which shows that the stainless steel coating has conferred substantial corrosion resistance. This indicates that the coating acts as a substantial barrier between the NaCl solution and the steel substrate, and increasing the coating thickness raises the effectiveness of the barrier.

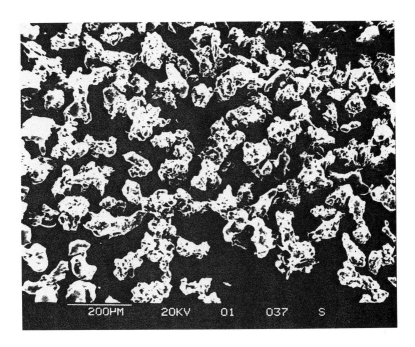

Figure 1 *Scanning electron micrograph of stainless steel feedstock powder*

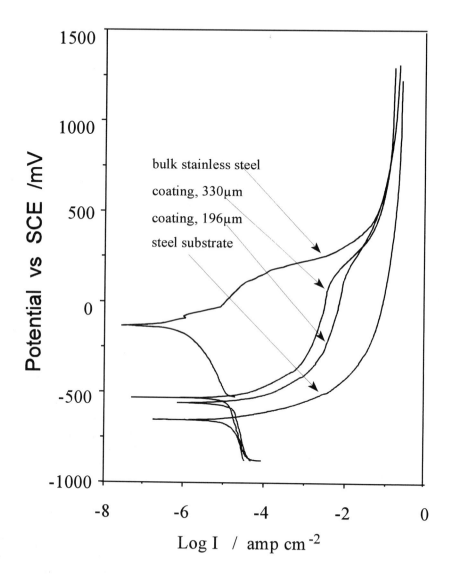

Figure 2 *Polarization curves of bare substrate, bulk stainless steel of same composition as feedstock powder and coatings of two thicknesses in tap water at 25°C*

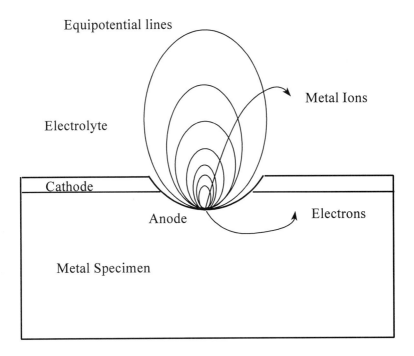

Figure 3 *Current flow and electric field associated with an electrochemically active site on a coated metal within an electrolyte*

The corrosion process was investigated in more detail using the scanning reference electrode technique, which detects local electrochemical activity over the surface of a specimen. A fine tipped probe capable of detecting extremely small variations in potential, is scanned over the specimen surface[3]. The corrosion process consists of a flow of metal ions into the solution above the corrosion site and a corresponding flow of electrons within the metal. The miniature current dipole sets up an associated electric field distribution within the electrolyte, which may be detected and quantified by the SRET (Figure 3).

SRET experiments were carried out on stainless steel coatings on steel in tap water without an applied DC potential. Under such test conditions, the stainless steel coating does not corrode significantly and the electrochemical activity detected is due to the corrosion of the underlying plain carbon steel substrate. The extent of electrochemical activity is indicated by colour contours as shown in Figure 4a. Figure 4a shows the results for a sample taken every 30 minutes in order to show the progress of corrosion. The maps clearly show that corrosion is localized: an area in the centre of the field is observed to gradually develop into a large anodic site. Figure 4b gives the corresponding potential variations of these anodic sites in terms of line scanning. Other anodic sites are also seen to initiate and grow at later stages.

A number of anodic sites identified by SRET were sectioned and observed in the scanning electron microscope. The observations revealed, in all cases, cross-sections of coating with

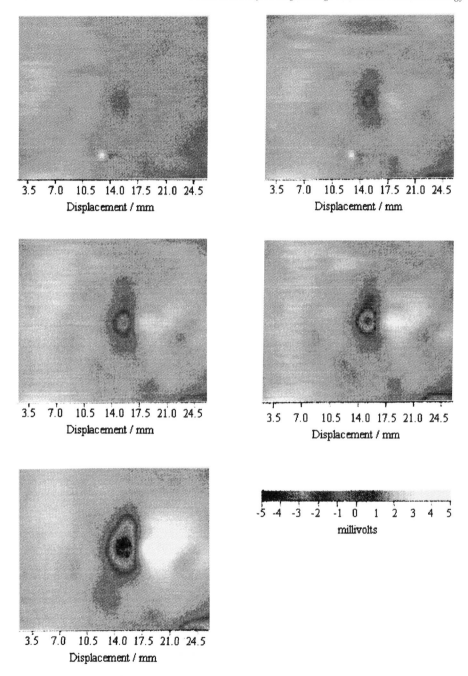

Figure 4a *Scanning reference electrode technique (SRET) maps of a stainless steel coated sample in tap water at 25°C recorded at 30 minute intervals (starting at 30 minutes at top left, 60 minutes top right, etc.)*

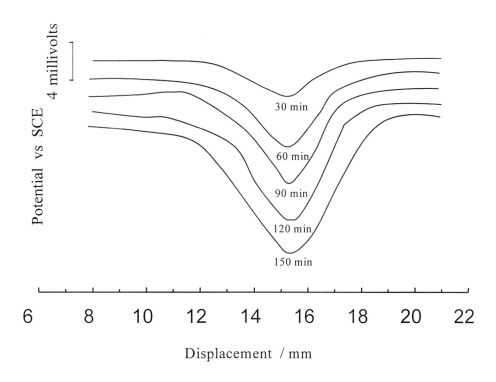

Figure 4b *Scanning reference electrode technique (SRET) linescans of a stainless steel coated sample in tap water at 25°C recorded at varying time intervals*

unusually high incidence of large pores as shown in Figure 5. In addition, significant separations between the coating and the substrate were observed at the interface below the active sites. Those observations suggest that corrosion is initiated at pores in the coating that form connective paths down to the interface, where anodic dissolution takes place. The microscopy also suggests that the pores form as a result of inadequate bonding between neighbouring splats or unmelted particles.

Figure 2 indicates that the corrosion rates of the coated steel decreases as the coating thickness is increased from 195μm to 330μm. Further work was carried out on the effect of coating thickness on corrosion and the data shown in Figure 6 indicate that the corrosion rate of the coated steel decreases markedly with increasing thickness. For example, the corrosion current in the NaCl solution decreases from $26\mu A\ cm^{-2}$ at 140μm thickness to $9\mu Acm^{-2}$ at 340μm thickness (Figure 6a); this compares with $130\mu Acm^{-2}$ for the bare substrate and $1\mu Acm^{-2}$ for bulk stainless steel. Similar trends were found with the hydrochloric acid solution(Figure 6b).

The effect of the coating thickness was investigated further using SRET. Figure 7 gives the data obtained for samples of coating thickness of 110μm and 360μm. The 110μm coating

(a)

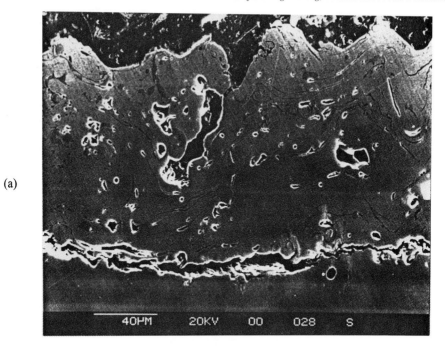

(b)

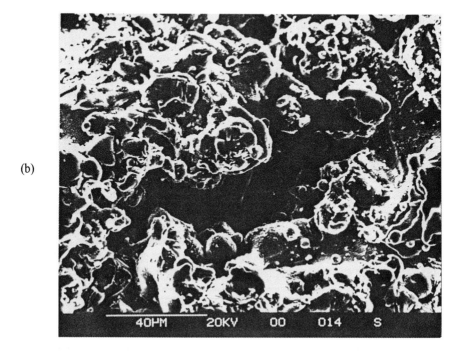

Figure 5 *Scanning electron micrographs of anodic sites detected by SRET on stainless steel coatings: (a) polished cross-section; (b) top surface*

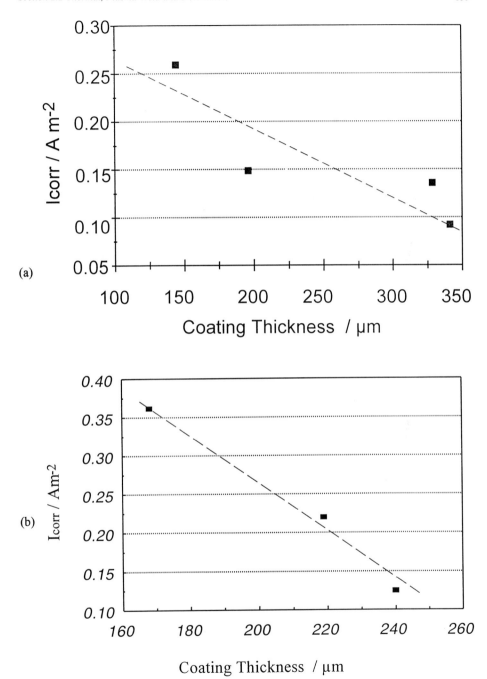

Figure 6 *Effect of stainless steel coating thickness on the corrosion rate in (a) 3.5wt% NaCl aqueous solution at 25°C, and (b) 2×10⁻³M HCl aqueous solution at 25°C – see also Tables 1 and 2*

Table 1 *Corrosion data of samples in 3.5wt% NaCl solution at 25°C*

Sample	Thickness / μm	E_{corr} / mV	I_{corr} / Acm^{-2}
Bare substrate	6000	–655	1.2999×10^{-4}
Bulk stainless steel	–	–122	1.053×10^{-6}
Coating 1	144	–448	2.5929×10^{-5}
Coating 2	196	–442	1.4873×10^{-5}
Coating 3	330	–390	1.4265×10^{-5}
Coating 4	341	–650	9.225×10^{-6}

showed a major anodic site on the map and line-scan, whereas no localized anodes appeared on the 360μm coating which suggests that there were no electrochemically active sites on this surface. The small, apparently cathodic features as resolved on the line-scan data were small bubbles on the surface of the specimen; the distribution of the local positive signals matched exactly with the observed distribution of these bubbles. The results suggest that the reduction in the corrosion rate with increasing thickness is due to a decrease in the incidence of through-thickness or connective pores.

The corrosion currents were measured as a function of time using chronoamperometry (CA). The data in Figure 8 show that the corrosion rate of a 160μm thick coating increases rapidly and then reaches a constant value. Similar trends were found for 220μm and 400μm coatings. The CA tests were repeated for the 220μm coating at elevated temperatures of the tap water (Figure 9). A similar trend was found at 60°C but a continual increase in corrosion rate was observed at 90°C.

The results show that corrosion in the plasma sprayed coatings initiates on a microscopic scale. Sectioning of anodic sites has revealed that pores are associated with the electrochemical activity and are considered to be the enabling mechanism for corrosion. The solution permeates the coating through the pores and reacts with the substrate at the interface. The iron from the plain carbon steel substrate dissolves in the neutral aerated solution to form Fe^{2+} by the anodic reaction:

$$Fe \rightarrow Fe^{2+} + 2e^{-}$$
{1}

Table 2 *Corrosion data of samples in $2 \times 10^{-3}M$ HCl at 25°C*

Sample	Thickness / μm	E_{corr} / mV	I_{corr} / Acm^{-2}
Bare substrate	6000	–641	1.6646×10^{-4}
Bulk stainless steel	–	+27	5.736×10^{-6}
Coating 5	168	–441	3.6208×20^{-5}
Coating 6	219	–483	2.2176×10^{-5}
Coating 7	240	–530	1.2451×10^{-5}

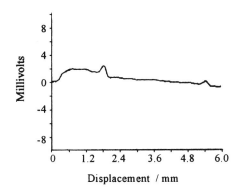
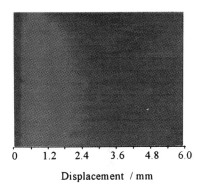

(a) 360μm

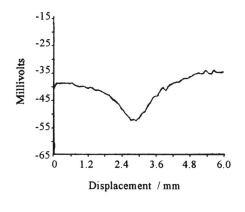
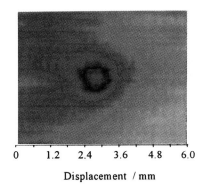

(b) 110μm

Figure 7 *SRET maps and line-scans of stainless steel coatings after 2.5 hours in tap water at 25°C with coating thickness of: (a) 360μm and (b) 110μm*

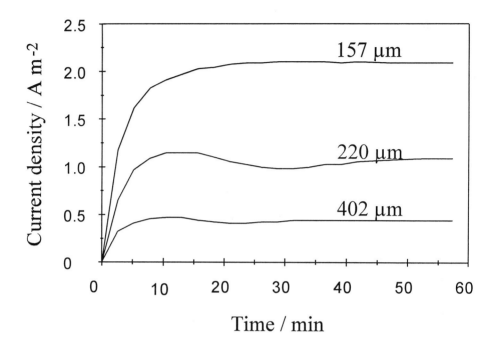

Figure 8 *Chronoamperometric curves for different thickness coatings in tap water at 25°C*

while the dissolved oxygen in the water is reduced by the cathodic reaction:

$$O_2 + 2H_2O + 4e^- \rightarrow 4OH^-$$ {2}

The dissolved ferrous and hydroxide ions then combine with further oxygen and water molecules to form hydrated ferric oxide or rust.

The chronoamperometry tests in Figure 8 show a two-stage behaviour: a first stage of increasing corrosion with time followed by a second stage of approximately constant corrosion rate. The SRET tests showed that the active sites initiated throughout testing. The first corrosion stage is thus taken to correspond to the period required to saturate all the connective pores with the solution. The solution first flows rapidly through the large pores to reach the interface but more time is required to permeate the narrower pores. More pores gradually become filled with solution and so the area fraction of substrate at the bottom of the pores that is in contact with the solution progressively increases with time. To a first approximation, the current density of the coated substrate i_c is given by:

$$i_c = i_s \, p$$ {3}

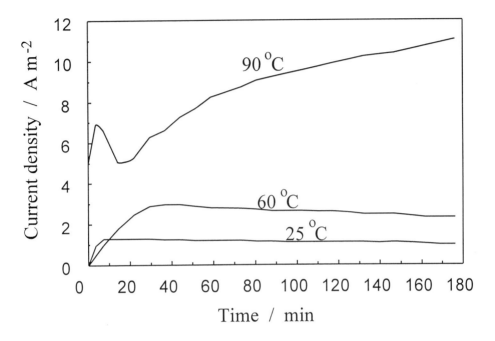

Figure 9 *Chronoamperometric curves for a 220μm thick coating in tap water at temperatures of 25°C, 60°C and 90°C*

where i_s is the current density of the bare, uncoated substrate and p is the area fraction of the substrate connected to the solution by porosity. A progressive increase in p is expected to lead to a monotonic increase in the corrosion rate, which is consistent with the first stage as observed in Figure 8. When all the connective pores are saturated with the solution, no further increases in corrosion rate occurs as shown in Figure 8.

The fact that the current density reaches a limiting value at high potentials (Figure 2) is attributed to concentration polarization and, in particular, the shortage of oxygen in solution to progress reaction {2}. This is in broad agreement with the absence of a passivation stage in Figure 2, which is also attributed to insufficient dissolved oxygen to generate a passive film.

The reduction in corrosion rate with increasing coating thickness (Figure 6) is considered to originate partly from the fact that the pore size is generally related to the feedstock particle size and is independent of the coating thickness. An unmelted particle undergoes little flow on impact with the substrate which results in the formation of surrounding voids of a size that is related to that of the original particle. The size of the voids formed by the inadequate bonding of one splat to another will also be related to the original feedstock particle size (Figure 5). The intersection of voids can then lead to a continuous path from the top surface of the coating down to the substrate at the interface. However, for a fixed void size, the probability of forming a connective path to the substrate interface will reduce with increasing coating thickness. The area fraction of substrate in contact with the solution through this connective porosity is therefore expected to decrease with increasing coating thickness. A further factor is the longer diffusion distance for the solution to the substrate as the coating becomes thicker. These effects are in accordance with the observed results (Figure 6).The increase in corrosion rate with temperature as observed in Figure 9 is attributed to the increased diffusivity of oxygen in water and the enhanced reaction rate at the substrate interface.

In the cases observed, the porosity was attributed to unmelted particles and inadequate bonding between splats. Particle melting depends upon particle size and the plasma spraying operating conditions, such as arc power, while intersplat bonding depends on factors, such as residual stress. Measures to control the particle size, operating conditions and residual stress of stainless steel coatings are thus expected to raise their corrosion resistance. An alternative approach to improve the corrosion resistance is to block the connective porosity of the coatings by the use of an undercoat between the coating and the substrate or a sealing coat on the top surface of the coating. Both of these measures may be achieved by the use of a dual powder feed system on the plasma spray equipment, which enables two different types of feedstock powder to be injected sequentially into the jet.

4 CONCLUSIONS

1. Scanning reference electrode and electron microscopy analyses of plasma sprayed stainless steel coatings on plain carbon steel indicate that corrosion occurs locally at pores within the coating.
2. Corrosion initiates first at the larger pores and later develops at smaller pores connected to the substrate.
3. Chronoamperometry shows that the corrosion rate initially increases with time and, when all the connective pores are saturated with solution, attains a constant value.
4. The corrosion rate decreases with increasing coating thickness in tap water, NaCl and hydrochloric acid solutions. This is attributed to the reduction in connective porosity and longer solution diffusion distances in thicker coatings.
5. Limiting current densities were observed for all stainless steel coatings in tap water due to the low concentrations of dissolved oxygen in solution. The corrosion rate increases markedly as the temperature is raised over the range 25°C to 90°C and this is attributed to the increased diffusivity of oxygen in water and enhanced reaction rates at the substrate interface.

Acknowlegements

The work was sponsored by the Government of the People's Republic of China and the British Council. The authors would like to thank the above for permission to publish the paper. The authors are also grateful to Mr M. Dawson of EG&G (UK) Ltd for his help in obtaining the SRET data.

References

1. G. Daufin, *Corrosion*, 1985, **41**, 533.
2. D. A. Sargeant, J. G. C. Hainse and S. Bates, *Materials and Technology*, 1989, **5**, 487.
3. H. S. Isaacs and V. Brijesh, 'Scanning reference electrode techniques in localized corrosion', Electrochemical Corrosion Testing, ASTM STP 727, Florian Mansfeld and Ugo Bertocci, Eds., American Society for Testing and Materials, 1981, p. 3.
4. 'Standard reference method for making potentiostatic and potentiodynamic anodic polarization measurements', ASTM G5-82, American Society for Testing and Materials, 1982, p. 122.

2.2.8
Thin Chromium Coatings Processed by HVOF-Spraying

E. Lugscheider, P. Remer and H. Reymann

MATERIALS SCIENCE INSTITUTE, UNIVERSITY OF TECHNOLOGY AACHEN, JNLICHERSTR. 342-352, 52070 AACHEN, GERMANY

1 INTRODUCTION

Within recent years high velocity oxygen fuel (HVOF) spraying has been considered an asset to the family of thermal spray processes[1]. Especially for thermal spraying of materials with melting points below 3000K it has proven successful, since it shows advantages in density and bond strength making it attractive for many wear and corrosion resistant applications[2]. Its high coating quality results from the use of a hot, combustion-driven high-speed gas-jet for thermal spraying[3].

The field of thermal spraying applications can be enlarged into regions of thin coatings, thicknesses ranging from about 50 to 120μm, especially using the HVOF process seems to be advantageous. In this area galvanic hard chromium platings have a great field of application. These platings are applied to reduce friction and corrosion, but imply several disadvantages. During their production they show environmental disadvantages, as the toxic behaviour of the solvents and reactants. HVOF-coatings applied with the CDS- and the Diamond Jet system are investigated for the possibility of competing with these galvanic platings.

2 EXPERIMENTAL PROCEDURE

Coatings using different grain size fractions of chromium powders were sprayed onto steel substrates which were prior grit blasted with pure Al_2O_3 (F24, 0.6– 0,85 mm). Two different nozzles for the CDS torch were used. Some information on the powder qualities is given in Table 1.

The CDS coatings were produced using the 4"- and 5"-nozzle. The combustion of the process gases is internal at elevated pressure and the variation of the length of the expansion nozzle offers the possibility to vary the dwell time of the particles and the particle velocity.

Table 1 *Specific data of the powders used*

	Powder 1	**Powder 2**	**Powder 3**
Particle form	Irregular, blocky	Irregular, blocky	Irregular, blocky
Particle size, nominal	+5.6–25 μm	+25–32 μm	+5.6–32μm

Especially for the gas flow rates of C_3H_8, O_2 process gases and N_2 powder carrier gas, the spraying distance, the powder feed rate and traverse speed of the torch relative to the substrate parameter variations were performed. These variations should give the parameters of main influence on the coating properties.

To make comparable coating qualities with different processes the Diamond Jet process was choosen. This process has an external combustion. Here the high particle velocities are reached by higher flow rates of the process gases. An overview on the spraying conditions is given in Table 2.

The evaluation of the coating quality in comparison to the galvanic platings is based on wear and corrosion tests, as well as metallographic investigations using microscopy, SEM and microprobe.

X-ray diffraction is used for the determination of the different phases that are generated during thermal spraying of the pure chromium powder. Monochromatic Co-K_α radiation with wave length of $\lambda = 1.7902 \times 10^{-10}$ m in combination with the equation of Bragg:

$$n\lambda = 2D\sin\theta \text{ for } n = 1, 2, 3 \dots \qquad \{1\}$$

where D is the crystallytic distance, θ the angle enclosed by an incoming and reflected X-ray beam, and λ the wave length. Using the ASTM-data the measured peaks can be evaluated.

For the evaluation of the wear behaviour of the coatings two standard test rigs were used. The pin-on-disk apparatus, with SiC–paper used as counterbody, gives information on the wear resistance against abrasive wear of the investigated coatings. The experimental setup is given in Figure 1.

Wear samples for pin-on-disk tests are pins of 14 mm in diameter and length of 38 mm coated on the front side. The load was 500 g on each pin, the SiC–paper was 400 grid. Sliding velocity was selected to 09.36 m/s and during the tests the pins were rotated to get constant wear on the whole surface of the pin and wear rates were determined by weight loss of the specimen. Water was used as lubricant.

The second wear test used was the Taber Abraser, giving information on abrasive-rolling and sliding wear. The main mechanism of wear is abrasion. The coatings were applied to

Table 2 *Spraying conditions*

Parameter	CDS	Diamond Jet	Diamond Jet
Propane (SLPM)	45–90	2800–3890	Ethene, 4700
O_2 (SLPM)	420–480	12000–16700	11200
N_2 (SLPM)	25–30	135	140
Spray distance (mm)	200–320	150–250	150–200
Powder feed rate (g/min)	10–18	48–55.5	48
Traverse speed (mm/sec)	100	500	500
Nozzle	4", 5"		

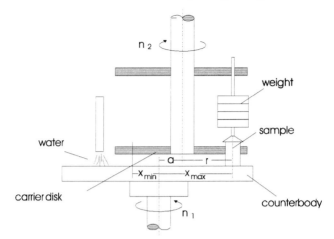

Figure 1 *Pin-on-disk apparatus*

disk-shaped specimens which are mounted onto a disk and two SiC-rolls move on these disks. The rolls are loaded with 1 kg each, the disk rotary speed was 60 rpm and the weight loss of the disks was measured every 250 revolutions. For the setup of the test rig, see Figure 2. Corrosion resistance of the different coatings and the galvanic plating was evaluated by potential measurements in different acids. Parameters are given in Table 3.

3 RESULTS AND DISCUSSION

3.1 Coating Microstructure

The light microscopic investigations into the microstructure of the coatings should give an insight in the phases that were created during spraying, the interfacial structure between substrate and coating and the porosity. Porosity of the coatings was evaluated using an

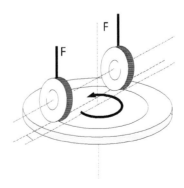

Figure 2 *Taber Abraser wear test*

Table 3 *Parameters of the corrosion tests*

Parameter	Settings
Medium	HCl, H_2SO_4, NaOH
Velocity of change of potential dU/dt	12mV/min
Electrode	$Hg/Hg_2Cl_2/Cl$
Difference of potential from electrode to standard H_2-electrode	+ 242mV
Maximum potential	2V

interactive picture analysis system by KONTRON. Depending on the powder used the porosity showed rather wide variation – the porosity varies in the range of 1 to 6 %. It is obvious that powder 3 gives higher porosity due to the wider grain size distribution.

The sprayed coatings show the typical build-up from numerous lamellae having different colours which are visible in the cross-sections of the samples. All coatings consist of light and dark phases with a higher amount of darker phases in the substrate near region. The higher porosity in the coatings of *powder 3* is mainly influenced by the wide grain size distribution of this powder. Big unmolten particles can be seen in the coating microstructure. The variation of the parameters had no significant influence on the coating growth.

For comparison to the HVOF-sprayed coatings Figure 3 gives a cross-section of a galvanic plating with a thickness of approximately 80μm. This galvanic plating shows a dense coating with good bonding to the substrate and without pores or any enclosures. The plating is built out of two layers, the deposition process was interrupted for measurement of coating thickness. Microcracks can easily be seen. These cracks influence the corosion behaviour of the plating

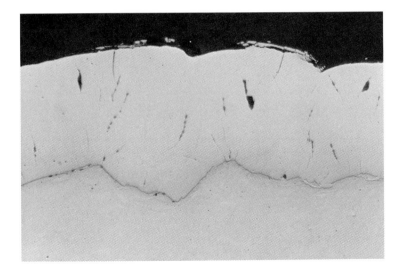

Figure 3 *Cross section of a galvanic coating*

and a notch effect on the substrate may arise from there. A disadvantage of the galvanic process is the dependence of the platings thickness on the geometry of the coated parts[4,5]. Therefore more work has to be focussed on grinding and finishing of the coated components.

Figures 4 and 5 show examples of coatings sprayed with the different nozzles of the CDS-System, Figure 4 giving an example of a sample sprayed with the 4"-nozzle. This coating has a porosity of approx. 1%. The microstructure shows several darker phases and unmolten particles can be seen.

In Figure 5 the cross-section of a sample sprayed with the 5"-nozzle is given. The microstructure is more regular and the lamellae are more flattened due to the higher degree of melting of the powder particles. This can easily be explained by the longer duration the particles remain in the hot region of the flame and the higher particle velocity reached. The porosity is slightly lower than in the 4"-coating, below 1% measured by image analysis.

Comparing the microstructure of the CDS- and Diamond Jet-sprayed coatings – Figure 6 – it is obvious that a lot more unmolten or partly molten particles occur in the Diamond Jet process. Due to the external combustion of the gases the dwell time of the particles in the hot zone is shorter and more kinetic energy is needed to create dense coatings. So the content of pure chromium phases increases, but the porosity also.

3.2 Phase Analysis

The different phases visible in the cross-sections of the coatings were analysed by X-ray diffraction. The structure of the coatings consists of phases from pure chromium, as well as the oxides Cr_2O_3 and Cr_3O_4. Most area is covered by pure chromium, followed by the Cr_2O_3. However, Cr_3O_4 comprises only little part on the structure. In all a ratio of chromium-to-chromium oxides is approximately 5:1. The unmolten particles and the light phases in the microstructure represent the pure chromium, while the darker phases consist of the two oxides.

Figure 4 *Cross section of a CDS sprayed coating, 4"-nozzle*

Figure 5 *Cross section of a CDS sprayed coating, 5"-nozzle, 500 X*

The generation of the chromium oxide phases is based on the high temperatures during processing and the relatively long period the particles are in flight surrounded by the hot gas-jet. During this time the molten, highly reactive particles react with the oxygen left from the combustion process, forming the phases Cr_2O_3 and Cr_3O_4.

The investigations showed that only the chromium-to-chromium oxide ratio is changing while the phases remain stable. This reaction occuring in flight can be influenced by the choice

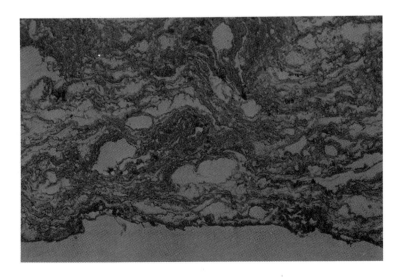

Figure 6 *Cross section of a Diamond Jet sprayed coating*

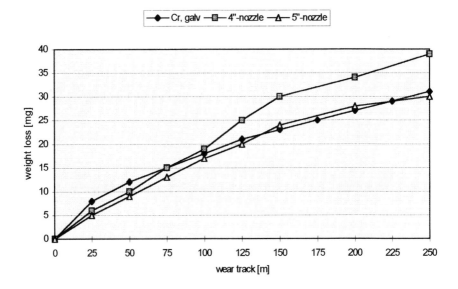

Figure 7 *Pin-on-disk wear behaviour of the coatings*

of parameters. Reducing the oxygen in the process gas ratio leads to a lower degree of oxidation in the coating.

3.3 Wear Properties

The median wear rate of four pins has been taken for evaluation. As can be seen in Figure 7, the wear rate of the sprayed chromium coatings shows a constant, linear wear rate, while the behaviour of the galvanic coatings is lower than the 4"-nozzle sprayed coating. The relatively soft chromium plating has a high wear rate in the beginning because the higher roughness reduces the area of contact under load. This process seems to be finished after the first 50 m of wear track.

The wear rates of the Taber Abraser wear test – Figure 8 – show similar results to the pin-on-disk ones and the coatings with a low content of unmolten particles show best results. One main difference between the two wear tests has to be mentioned, the SiC-paper at the pin-on-disk test is changed every cycle and the counter rolls of the Taber Abraser are cleaned by pressurized air. The hard chromium plating tends especially to smear the rolls and a quasi-static wear behaviour develops because of the progressively increasing similarity of the wear partners. *Powder 3* generally shows lower wear resistance due to the above mentioned reasons.

Regarding an initial phase of approx. 2000 m wear track the results of the sprayed coatings show a lower increase in weight loss than the galvanic coating. This initial wear is due to the coating's "as-sprayed roughness", the roughness is flattened and then the wear track has the same dimensions as that of the galvanic coating. The wear behaviour of the sprayed coatings is favourable to that of galvanic coatings. This results from the oxide phases in the coatings which have higher hardness and are embedded in the pure chromium particle matrix.

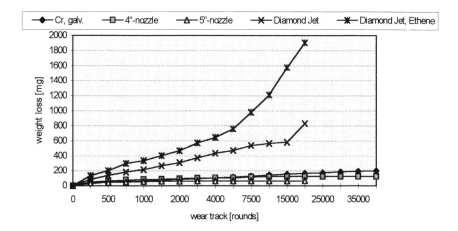

Figure 8 *Taber Abraser wear behaviour of the coatings compared to galvanic coating*

3.4 Corrosion Properties

The coating structure influences strongly the corrosion behaviour of the coatings. If the porosity is high, the corrosive medium can attack the substrate very easily. So the corrosion resistance of HVOF sprayed coatings is influenced intensively by the build-up of the coatings.

Depending on the corrosive medium the behaviour of the sprayed coatings shows more or less significant differences to the galvanic plating. Exposed to H_2SO_4 the HVOF-coatings show passivation at low level of potential which reaches a steady state early and the break through potential of the coatings is similar to that of the galvanic platings. In HNO_3 the sprayed coatings showed a very fast rise in the intensity, so that the corrosion behaviour in this medium is not sufficient. Looking at NaOH the coating passivates at the beginning and no significant difference in the potential can be seen compared to the galvanic coatings.

In contrast to the galvanic coatings which have a wide passive area, the sprayed coatings imply an ongoing oxidation at low level. This is initiated due to the oxide phases in these coatings. Only coatings which have high porosity or other defects show a higher level of corrosion attack. According to the position of these defects the corrosion takes place on the surface, in or under the coating, as can be seen in, Figure 9.

In the case of corrosive attack it can be stated that dense coatings with a good bonding to the substrate show a corrosion resistace which is similar to that of galvanic coatings.

4 CONCLUSIONS

In the current investigations it could be proven that thermally sprayed coatings from pure chromium powder using the HVOF process obtain similar properties to galvanic coatings. The sprayed coatings have a very low porosity and coating thickness is adjustable in the range of 50 up to 250μm. Due to the thermal spray process the coatings do not consist of pure

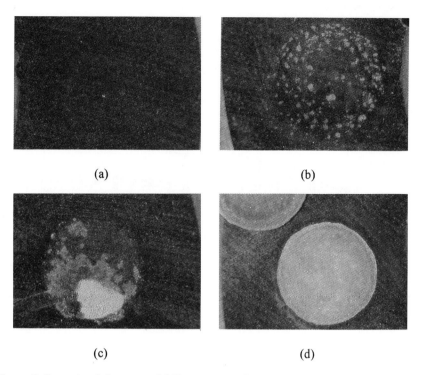

(a)   (b)

(c)   (d)

Figure 9 *Corrosion behaviour of different sprayed coatings: (a) no corrosion attack visible, (b) corrosion on the surface, (c)corrosion inside the coating and (d) corrosion under the coating*

chromium. During the relatively long lasting period of flight the particles react with oxygen left from the combustion process. Chromium oxides Cr_2O_3 and Cr_3O_4 are generated. The lamellar structure of the coatings shows a mixture of these phases and the ratio between the chromium and the chromium oxide phases is approx. 5:1.

Both process variants, the CDS-process and the Diamond Jet process gave good and comparable results. Those coatings having low porosity, good melting behaviour of the particles and a fine distribution of phases show a similar behaviour regarding wear and corrosion at galvanic platings.

It has to taken into account that best coatings could be sprayed with powder [2], which had a very narrow grain size distribution. The degree of melting of the powder particles is of main interest especially for the creation of thin, dense coatings. Only in the case of fully melting the particles, coatings can be achieved which have properties similar to those of galvanic chromium platings.

Looking at the environmental aspects of the two processes, advantages can be seen for the thermal spray process. Thus new fields of application can be explored for the thermal spray technology. Especially for coating of rotional parts of median size the relativly high deposition rate of the HVOF process offers advantages for a more economical and ecological coating solution.

Acknowledgements

The work was carried out as part of the joint industrial research promoted by the Federal Minister of Economic Affairs through the Study Group of Industrial Research Associations with the support of the German Welding Society. Our thanks are due to this support. The authors thank Dr.-Ing. H. Reimann, Gotek GmbH, Dipl.-Ing. P. Heinrich, Linde AG, Dr.-Ing. Th. Weber, Deloro Stellite, Dr.-Ing. G. Kalawrytinos, Pallas GmbH and Dipl.-Ing. E. Schwarz, Bauhammer Metallspritz GmbH for a lot of fruitful discussion during this project.

References

1. M. L. Thorpe, Thermal Spray Industry in Transition, *Advanced Materials & Processes*, May 1993.
2. J.R. Moens, G. Barbezat, A.R. Nicoll, Eigenschaften und Anwendungen von CDS-Schichten, Technische Rundschau Sulzer, 2/1991.
3. C.M. Hackett, G.S. Settles, J.D. Miller, 'On the gas dynamics of HVOF Thermal Sprays', Proc. of the 1993 Nat. Therm. Spray Conf., Anaheim, CA, 7-11 June 1993.
4. Lehrbuch der Galvanotechnik, Praktische Galvanotechnik, 4. Auflage, 1984, Eugen G. Leuze Verlag.
5. Dettner/Elze, Handbuch der Galvanotechnik, 1969, Carl Hanser Verlag.

2.2.9

Dilution Control in Weld Surfacing Applications

J.N. DuPont and A.R. Marder

DEPARTMENT OF MATERIALS SCIENCE AND ENGINEERING, LEHIGH UNIVERSITY, BETHLEHEM, PA, USA

1 INTRODUCTION

The selection of processing parameters for surfacing applications is a difficult task for personnel involved in welding procedure development. There is a large number of parameters which must be optimized to ensure the deposition of reliable claddings in an efficient manner. For example, achieving a low dilution level is an important objective of a surfacing procedure development program. Low dilution of the cladding with the substrate (most commonly steel) is desirable to preserve wear and corrosion resistance. In general, low dilution is accomplished by the use of high filler metal feed rates at the lowest possible energy delivered by the arc. The use of low arc energy holds the degree of substrate melting and concomitant dilution to a minimum. However, if the energy is too low for a given filler metal feed rate, then a large portion of the energy is consumed in melting the filler metal with little energy remaining to melt the substrate. As a result, the cladding does not completely fuse to the substrate and is susceptible to spallation during service. This transition is depicted schematically in Figure 1 which shows the change in dilution and fusion which occurs as the filler metal feed rate is increased for a given level of energy delivered by the arc.

In Figure 1a, the filler metal feed rate/arc energy ratio is low. In this situation, only a small portion of the total energy is required for melting the filler wire. The remaining energy is consumed by melting a rather large portion of the substrate. Although there is complete fusion of the cladding with the substrate, the level of dilution is high due to the large quantity of substrate melted. As the filler metal feed rate is increased, Figure 1b and 1c, a larger portion of the total energy is required for melting the filler wire and less energy is available for melting the substrate. As a result, the quantity of substrate melted decreases. Figure 1c represents the optimum balance of parameters because the highest possible feed rate is achieved for a given level of energy while maintaining complete substrate/cladding fusion and a low level of dilution. With a further increase in filler metal feed rate, Figure 1d, the filler metal absorbs a major portion of the arc energy with insufficient energy available to completely melt the substrate. At this point, complete cladding/substrate fusion is lost and the critical filler metal feed rate/arc energy balance is considered to be exceeded.

The problem discussed above is especially applicable to the non-consumable electrode processes where the arc current and filler metal feed rate are independently controlled. With the consumable electrode processes, the characteristics of the power source are such that an increase in wire feed speed automatically provides an increase in arc current (and hence arc

CONSTANT ARC POWER, $\eta_a VI$

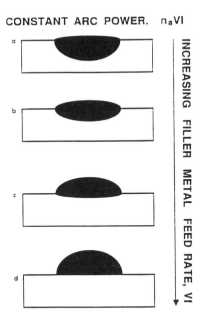

Figure 1 *Changes in dilution and fusion which are expected to result as the filler metal feed rate is increased for a given level of energy delivered by the arc*

energy). However, the incremental increase in current with wire feed speed can be affected by electrode extension, arc voltage, and power source slope. As such, the critical balance of parameters can also be difficult to obtain with the consumable electrode processes.

It is clear, then, that the process parameters must be carefully balanced to provide low levels of dilution while maintaining complete fusion of the cladding with the substrate. However, despite the extensive use of welding as a means of applying protective claddings, a method for predicting this critical balance is presently not available. As a result, procedure development is currently conducted using time consuming, iterative approaches and the results are generally only applicable for a given application. Thus, a quantitative method for predicting low dilution would be of significant value for surfacing applications. Such an approach is proposed here for the simplest case of a single pass weld as shown schematically in Figure 2 where the Dilution, D is given by:

$$\% Dilution = \frac{A_s}{A_s + A_{fm}} \times 100 \qquad \{1\}$$

where A_s is the melted cross-sectional area of the substrate and A_{fm} is the cross-sectional area of the deposited filler metal.

Deposited filler metal cross-sectional area, Afm

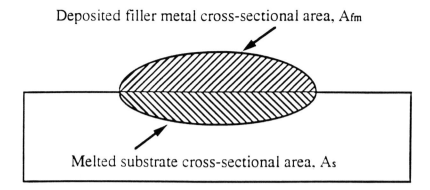

Melted substrate cross-sectional area, As

Figure 2 *Schematic illustration of dilution in single pass welds*

2 EXPERIMENTAL PROCEDURE

Estimation of dilution with the approach presented here requires knowledge of the thermal efficiency (arc efficiency and melting efficiency) of the welding process. The methods and equipment utilized for measuring these thermal efficiency factors have been discussed in detail in a separate papers[1]. The welding processes evaluated included PAW, GTAW, GMAW, and SAW. The thermal efficiency factors were determined under the same range of parameters used throughout the present discussion (Table 1).

A fully automated welding system designed specifically for research was used for all the experiments. A 500 A constant current/constant voltage power source was used for each process. A separate plasma console unit was used for control of the pilot arc, plasma gas, and shielding gas for the PAW process. Motion of the individual torches was provided by an automated travel carriage. The power source, travel carriage, and all auxiliary equipment are controlled by a Texas Instruments/Siemens programmable control unit.

To simulate the simple dilution condition illustrated schematically in Figure 2, single pass welds were prepared by depositing type 308 austenitic stainless steel onto 305 mm square by 6.4 mm thick A36 steel substrate under the range of parameters listed in Table 1. Each deposit was approximately 254 mm in length. The ranges listed in Table 1 for each process were determined by preliminary weld trials. The lower limit to travel speed for a given arc power was governed by the formation of excessively wide and deeply penetrating welds. The upper limit of travel speed was established for a given arc power when the process could no longer adequately melt the substrate and filler metal. Details of the individual processes are provided elsewhere[1].

After welding, each sample was cross-sectioned using an abrasive cut-off wheel, polished to a 1 pm finish using silicon carbide paper, and etched in a 2% Nital solution. The individual cross-sectional areas of the melted substrate (A_s) and deposited filler metal (A_{fm}) were then measured using a LECO quantitative image analysis system. The dilution was determined for each sample by Equation {1}.

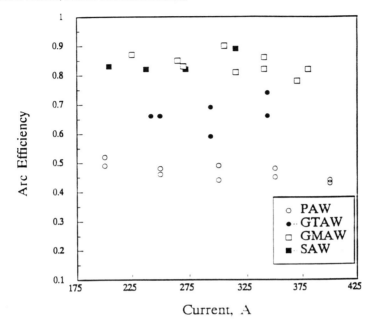

Figure 3 *Arc efficiency as a function of current for the Plasma Arc Welding (PAW), Gas Tungsten Arc Welding (GTAW), Gas Metal Arc Welding (GMAW) and Submerged Arc Welding (SAW) processes*

3 RESULTS AND DISCUSSION

3.1 Thermal Efficiency Factors and Dilution

As previously noted, estimation of dilution with the approach presented here requires knowledge of the thermal efficiency of the welding process. The term thermal efficiency describes the welding process in two ways, namely arc efficiency and melting efficiency. Arc efficiency is the fraction of total process energy which is actually delivered to the substrate and weld deposit. The ratio of energy used for melting to that which is delivered to the work-piece defines the melting efficiency.

Figure 3 shows the arc efficiency for each welding process considered as a function of welding current. These measurements were conducted using a Seebeck arc welding calorimeter

Table 1 *Experimental matrix of processing parameter ranges used in dilution and thermal efficiency measurements*

Process	Current (A)	Voltage (V)	Travel speed (mms^{-1})	Filler metal feed rate mm^{3}s^{-1}
PAW	250–400	25–32	2–4	8–120
GTAW	250–400	15–16	6–10	20–130
GMAW	230–400	27–36	6–26	120–245

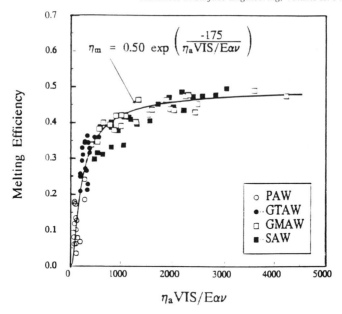

Figure 4 *Variation in melting efficiency with $\eta_a VIS/E\alpha v$ parameter for the Plasma Arc Welding (PAW), Gas Tungsten Arc Welding (GTAW), Gas Metal ArcWelding, (GMAW) and Submerged Arc Welding (SAW) process*

as first described for arc efficiency measurements by Giedt et all[2]. A clear distinction in the ability of each process to transfer energy to the work piece is evident in Figure 3. The data also show there is very little variation in arc efficiency for a given process over the current range investigated. The consumable electrode processes (GMAW and SAW) exhibit an average arc efficiency of 0.84 ± 0.04. The GTAW process has an average arc efficiency, of 0.67 ± 0.05, and the PAW process displays an average arc efficiency of 0.47 ± 0.03. These values are in good agreement with other arc efficiencies reported in the literature for these processes[3,4]. Thus, the arc efficiency for a given process under the broad range of current considered can be estimated reasonably well as a constant.

Unlike arc efficiency, the melting efficiency (η_m) depends strongly on the processing parameters, base metal properties, and base metal thickness. It has recently been shown that melting efficiency can be estimated from the welding parameters and material properties by an equation of the form[1,5].

$$\eta_m = Ae^{\frac{-B}{\eta_a VIS/E\alpha v}} \tag{2}$$

Where η_a is the arc efficiency, VI is the arc power, S is the travel speed, E is the enthalpy required for melting, α is the thermal diffusivity at 300 K and v is the kinematic viscosity at the melting point. The quantity ($\eta_a VIS/E\alpha$) is a dimensionless parameter. The constant A is representative of the maximum melting efficiency for a given joint design or substrate geometry which is obtained when the quantity ($\eta_a VIS/E\alpha$) is large. The constants A and B can be determined from the intercept (A) and slope (B) on a plot of $\ln(\eta_m)$ against ($\eta_a VIS/E\alpha$)$^{-1}$.

Figure 4 shows melting efficiency plotted against this dimensionless parameter for the present set of conditions. An average value between carbon steel and austenitic stainless steel of $E = 9.6$ Jmm$^{-3[6,7]}$ and $v = 0.84$ mm^{2}s$^{-1[8]}$ was used in the plot. The effect of thermal diffusivity should be controlled by the substrate, so the value for steel was used (9.1 mm^{2}s^{-1})[9]. With these values $E\alpha v = 73$ jmms^{-2}. Since the substrate and filler metal are held constant in this experiment and the value of $E\alpha v$ is therefore fixed, the plot of the data using this dimensionless parameter does not provide an opportunity to reveal the effectiveness of the parameter to normalize the influence of material property variations. However, it does reveal the proper dependence of melting efficiency on the process parameters (arc power and travel speed). In addition, this parameter has been shown to normalize differences in material thermo-physical properties between 304 stainless steel and NI 200 in an edge weld configuration[5]. The approach is adopted here to facilitate prediction of the melting efficiency in order to estimate the dilution. Equation {2} is plotted in Figure 4 with the constants $A = 0.50$ and $B = 175$, and the expression provides a reasonable representation of the experimental data. Thus, the arc efficiency is approximately constant for a given process under a broad range of current (Figure 3) while the melting efficiency can be estimated via Equation {2} where A and B are experimentally determined constants (Figure 4).

When dilution is determined by metallographic methods, it is typically calculated by the individual cross-sectional area terms of the deposited filler metal and melted substrate as given by Equation {1}. However, dilution is actually the result of the volumetric quantities v_{fm} and v_s where v_{fm} is the volume of deposited filler metal and v_s the volume of melted substrate. These volumetric terms are simply reduced by one dimension into an area term when the sample is cross-sectioned and the measurement of A_s and A_{fm} are made, the assumption being that the cross-sectional areas do not vary along the length of the sample. Constant cross sectional areas are produced along the weld length when the volumetric melting rate of the filler metal, V_{fm}, and the substrate, V_s are constant with travel speed. Under this assumption, dilution can also be expressed in terms of the volumetric melting rates of the substrate and the filler metal:

$$\% Dilution = \frac{V_s}{V_s + V_{fm}} \times 100 \qquad \{3\}$$

Assuming the set value of volumetric filler metal feed rate, V_{fm}, is equivalent to the actual deposited volumetric filler metal rate (i.e., filler metal losses due to spatter are negligible), then the volumetric melting rate of the filler metal is a controlled variable of the process. An expression for V, can be obtained by considering a simplified balance of power terms across the welding arc which is facilitated by the thermal efficiency factors.

$$\eta_a \eta_m VI = V_{fm} \left\{ \int C_p(T)dT + \Delta H_f \right\}_{fm} + V_s \left\{ \int C_p(T)dT + \Delta H_f \right\}_s \qquad \{4\}$$

The $\left\{ \int C_p(T)dT + \Delta H_f \right\}_{fm}$ and $\left\{ \int C_p(T)dT + \Delta H_f \right\}_s$, terms ($C_p$ is specific heat, ΔH_f is latent heat of fusion) represent the enthalpy change required to melt a given volume of filler metal and substrate. These terms can be represented by E_{fm} for the filler metal and E_s for the substrate for simplicity. The values for these thermo-physical properties under the present conditions are $E_{fm} = 8.7\,\mathrm{jmm^{-3}}$ [7] for type 308 stainless steel and $E_s = 10.5\,\mathrm{Jmm^{-3}}$ for carbon steel. The left side of Equation {4} represents the melting power delivered by the arc while the right side represents the power required for melting of the substrate and filler metal. Equation {4} can be rewritten in simpler form:

$$\eta_a\eta_m VI = V_{fm}E_{fm} + V_sE_s \qquad \{5\}$$

The arc and melting efficiencies can be estimated as discussed in the previous section. Thus, V_s, is the only unknown in Equation {5}. Rearranging {5}:

$$V_s = \frac{\eta_a\eta_m VI - V_{fm}E_{fm}}{E_s} \qquad \{6\}$$

Writing {3} in a more convenient form:

$$\%Dilution = 100.\left(1 + \frac{V_{fm}}{V_s}\right)^{-1} \qquad \{7\}$$

and substituting Equation {6} into {7}, an equation for dilution can be obtained:

$$\%Dilution = 100.\left(1 + \frac{V_{fm}E_s}{\eta_a\eta_m VI - E_{fm}V_{fm}}\right)^{-1} \qquad \{8\}$$

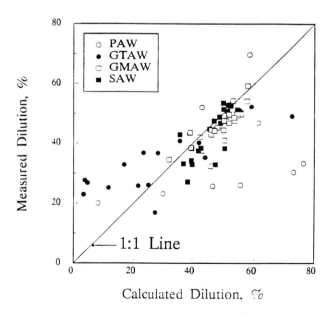

Figure 5 *Comparison of measured and calculated dilutions for each process evaluated*

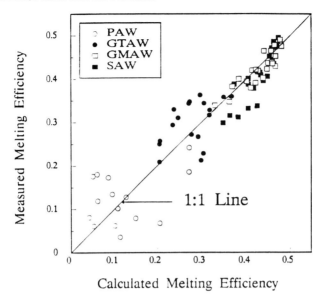

Figure 6 *Comparison of measured and calculated melting efficiencies for each process evaluated*

Equation {8} provides an analytical expression for dilution in terms of the welding parameters which are controlled during the process, thermal efficiency factors which can be estimated as previously discussed, and filler metal and substrate thermo-physical properties.

Figure 5 shows a comparison of measured and calculated dilutions for each process using Equation {8}, where the nominally constant values of arc efficiency for each process were used and melting efficiency values were determined by Equation {2} with $A = 0.50$ and $B = 175$. The agreement between experimentally measured and calculated dilution values shows a dependence on the process being compared. The agreement between measured and calculated dilution for the GMAW and SAW processes is reasonable while the correlation begins to breakdown for the GTAW process. The correlation is poor for the PAW process.

This behaviour results from the difficulty in accurately predicting melting efficiency at low values of $\eta_a VIS$ due to the exponential relation between η_m, and $\eta_a VIS$. This is displayed in Figure 6 where the calculated melting efficiencies (using Equation {2} with $A = 0.50$ and $B = 175$) and measured melting efficiencies of each process are plotted. The PAW and GTAW processes operate at relatively low values of $\eta_a VIS$ under the present set of conditions[1]. In this region, small increases in arc power or travel speed promote large increases in the melting efficiency. As a result, melting efficiency is difficult to predict accurately from Equation {2} in this range. The poor correlation of calculated and experimentally measured dilution for the PAW process is a result of the inability to accurately predict melting efficiency in this operating range. The GTAW process operates at slightly higher values of net arc power and travel speed where the slope of the η_m versus ($\eta_a VIS/E_a v$) curve begins to decrease (Figure 4). Thus η_m can be predicted slightly better and the calculated dilution values are slightly closer to those observed experimentally. Finally, the GMAW and SAW processes operate at relatively high values of net arc power and travel speed where η_m, varies only slightly with ($\eta_a VIS/E_a v$) and can be predicted quite accurately by Equation {2}. Thus, the dilution values can also be predicted reasonably well.

3.2 Graphic Display of Processing Parameter Effects

In many applications it is often desirable to operate at high filler metal feed rates (deposition rates) for economic reasons while maintaining low dilution levels for corrosion resistance. In this regard, it is useful to utilize the dilution equation to develop a graphical display which reveals the effect of processing parameters on dilution and provides an indication of the maximum filler metal feed rate, $V_{fm,max}$ which can be achieved for a given level of melting power, $\eta_a \eta_m VI$.

Equation {8} is first solved in terms of the deposition rate, V_{fm}:

$$V_{fm} = \left(\frac{\gamma}{E_s + \gamma E_{fm}} \right) \eta_a \eta_m VI \qquad \{9\}$$

where

$$\gamma = \frac{1}{D} - 1 \qquad \{10\}$$

where D is the fraction (i.e., not percent) dilution. Equations {9} and {10} indicate that, for a given filler metal/substrate combination (i.e., E_{fm} and E_s fixed), a plot of filler metal feed rate (V_{fm}) against melting power ($\eta_a \eta_m VI$) will yield various slopes which depend only on the dilution. An indication of maximum filler metal feed rate for a given melting power can be considered to be reached when the dilution is reduced to zero and the filler metal does not adequately fuse to the substrate. This occurs when the volumetric melting rate of the substrate is zero, $V_s = 0$. By setting $V_s = 0$ in the simple power balance of Equation {5}, the condition of maximum filler metal feed rate can be identified

$$V_{fm,max} = \frac{1}{E_{fm}} \cdot \eta_a \eta_m VI \qquad \{11\}$$

Equation {11} defines the condition where the filler metal feed rate is increased to the point where all the melting power would have to be used for melting the filler wire, in which case no melting power would remain to melt the substrate. This is the condition when the dilution is reduced to an undesirable level of zero. This is an obvious over simplification of the problem since it does not consider the relative rate of energy transport from the process to the filler metal and substrate. The condition is used here only to bound the limit for the maximum filler metal feed rate. Any filler metal feed rate utilized for a fixed melting power greater than that defined by Equation {11} will certainly result in some incomplete melting of the substrate and/or filler metal.

The maximum filler metal feed rate defined here varies linearly with melting power by a line with slope ($1/E_{fm}$). Therefore, equations {9} and {11} can be plotted on one diagram to reveal the maximum filler metal feed rate for a given melting power and the resultant dilution which will result from a given filler metal feed rate/melting power ratio. This concept is shown in the diagram in Figure 7. The filler metal feed rate (V_{fm}) is plotted as a function of the melting power ($\eta_a \eta_m VI$) and the slopes which correspond to various calculated dilution levels

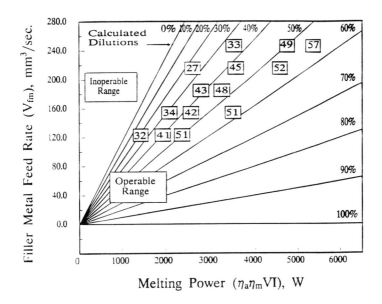

Figure 7 *Diagram showing effect of processing parameters on dilution with experimental data for SAW process plotted on the diagram*

are plotted in 10% increments. Since the filler metal/substrate combination is fixed in this experiment, the slopes are determined only by the dilution. A boundary between an "inoperable range" and "operable range", as defined by Equation {11}, is also plotted. This line is also denoted as the 0% dilution line. Data for the SAW process are plotted and the measured values of dilution are shown in the data points for comparison to the calculated iso-dilution lines. As with Figure 5, the agreement is reasonable for the SAW process. This is to be expected since Figure 7 is simply a re-plot of the data presented in Figure 5. However, with this diagram, the effect of the processing parameters on dilution is readily apparent. For a fixed filler metal feed rate, the dilution increases with increasing melting power. In this case, the extra melting power can not be absorbed by the filler metal if the filler metal feed rate is fixed, so the substrate absorbs the extra melting power which results in an increase in the volumetric melting rate of the substrate (V_s) and concomitant increase in dilution. Note that a filler metal feed rate of zero describes an autogenous weld which always has 100% dilution, as exhibited by the diagram. Conversely, for a given melting power, an increase in the filler metal feed rate results in a decrease in dilution. In this case, the filler metal consumes a larger portion of the total melting power and less energy is available to melt the substrate. As a result, the substrate volumetric melting rate decreases and dilution is reduced. When the filler metal feed rate is increased beyond the operable/inoperable boundary for a fixed melting power where Equation {11} is satisfied, the dilution reaches 0% (an undesirable condition). Thus, the effect of processing parameters on dilution and maximum filler metal feed rate are displayed on one diagram to facilitate parameter selection in applications where dilution and deposition rate are important.

The diagram also reveals the effect of differences in arc and melting efficiencies among the processes which leads to differences in the ability of each process to achieve high filler metal

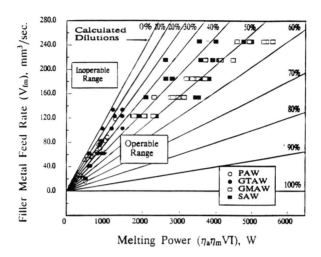

Figure 8 *Diagram showing effect of processing parameters on dilution with experimental data for all processes plotted on the diagram.*

feed rates under the present set of conditions. The consumable electrode processes exhibit the highest arc efficiency which, in turn, translates into high melting efficiency. By contrast, the non-consumable electrode processes exhibit lower arc and melting efficiencies. Under conditions of equivalent arc power (VI), the amount of energy delivered to the work piece for melting $(\eta_a\eta_m VI)$ is reduced when the arc and melting efficiency are lower. Considering Equation {11} as a measure of maximum filler metal feed rate it becomes apparent that high filler metal feed rates are favoured by high thermal efficiency. Therefore, to avoid 0% dilution, higher arc powers (VI) or lower filler metal feed rates will be required for processes with low thermal efficiency in order to compensate for these energy losses. In regard to these considerations, the diagram can also be used to aid in the selection of the most suitable process for a given application. This is accomplished by plotting the data for all the processes on the diagram and comparing the processes in terms of the maximum achievable deposition rate. This comparison is made in Figure 8 where all the data for the processes are plotted on one diagram and a distinct difference in terms of filler metal feed rate is observed between the consumable and non-consumable electrode processes. (The experimental dilution data for the GMAW and SAW processes will be close to the calculated iso-dilution values. However, the dilution correlation for the PAW and GTAW processes is poor for reasons previously discussed and is plotted here only to reveal differences in maximum achievable filler metal feed rate.) With the consumable electrode processes operated under the present set of conditions, the thermal efficiency is high which moves the operating conditions to high melting powers which, in turn, permits high deposition rates. The maximum deposition rate attainable in this work for the GMAW and SAW processes was 245 mm³/second. In contrast, the maximum deposition rate of non-consumable electrode processes (PAW and GTAW) was limited to 120–130 mm³/second because the thermal efficiency is low.

4 CONCLUSIONS

A study of dilution in single pass arc welds was conducted. Welds were deposited using type 308 austenitic stainless steel filler metal onto 6.4 mm thick A36 carbon steel substrate using the PAW, GTAW, GMAW, and SAW processes. The dilution which results from a given set of processing parameters (V, I, S, and V_{fm}) can be estimated by:

$$\% Dilution = 100.\left(1 + \frac{V_{fm}E_s}{\eta_a \eta_{fm}VI - E_{fm}V_{fm}}\right)^{-1}$$

where the arc efficiency, η_a, is nominally constant for a given process and melting efficiency, η_m can be estimated by:

$$\eta_m = Ae^{\frac{-B}{\eta_a VIS/E\alpha v}}$$

where A and B are experimentally determined constants. Under the present set of conditions, $A = 0.50$ and $B = 175$. Reasonable estimates of dilution are possible with these equations when the melting efficiency is relatively high and easy to accurately predict. The processing diagram proposed from the dilution equation facilitates selection of processing parameters in applications where dilution control is important.

References

1. J. N. DuPont and A. R. Marder, *Weld. J.*, 1995, **74**, 406.
2. W. H. Giedt, L. N. Tallerico, and P. W. Fuerschbach, *Weld. J.*, 1989, **68**, 28.
3. H. B. Smartt, J. A. Stewart, and C. J. Einerson: Proc. ASM Intl. Welding Congress, 1985, ASM 8511–01 1, p. 1.
4. A. D. Watkins, H. B. Smartt, and C.J. Einerson: *Proc. Recent*
5. P. W. Fuerschbach and G. A. Knorovsky: *Weld. J.*, 1991, **70** , 287.
6. H. A. Fine and G. H. Geiger, 'Handbook on Material and Energy Balance Calculations in Metallurgical Processes', AIME, Warrendale, PA, 1979, p. 431.
7. C. F. Lucks and H. W. Deem, 'Thermal Properties of Thirteen Metals', STP No. 227, ASTM, Philadelphia, PA. 1958.
8. E. A. Brandes, 'Smithells Metals Reference Book', 6th ed., Butterworths, London, England, 1983.
9. T. F. G. Grey, J. Spence, and T.H. North, 'Rational Welding Design', Newnes-Butterworths, London, England, 1975.

2.2.10
Wire Explosion Coating

M.G. Hocking

MATERIALS DEPARTMENT, IMPERIAL COLLEGE, LONDON, UK

1 INTRODUCTION

Internal coating of tubes is readily feasible, in air at atmospheric pressure, by wire explosion spraying. All metals, including refractory metals, can be sprayed by wire explosion. In correct usage, wire vapour reaches the substrate and drives out the air before impact of the following shower of metal droplets, Thus no steps are needed to exclude air, and coating occurs without oxidation. The droplets are spherical, usually about 2 or 3 microns diameter, and travel at about 600 m/s.

The substrate is unaffected by the process heat (paper can be coated) and glass and ceramics can be coated as easily as metals. About 60% of the weight of the exploded wire is deposited on the inner bore of a hollow cylinder if the wire is exploded axially in it.

The coating thickness from a single spraying is about 5 to 15 microns and repeated wire explosions are used to produce thicker coatings. The coating is more adherent and smoother than by plasma or flame spraying, due to the much higher droplet velocity and smaller droplet size; wire explosion spraying of Mo, W and steel wire onto Al and steel have 5 times better adhesion than by flame spraying[1]. Using the correct wire alloy gives any desired coating composition[1-4].

Wire explosion spraying is used industrially in the mass production of Al alloy cylinders for internal combustion engines, with Mo and other coatings. Different wire diameters can be used, from 10 to 1500 microns. Typically, a 2000 pF capacitor bank at 15 kV is required to explode a 7 mm length of 25 micron diameter wire. The current rises at about 10^{11} As^{-1} and current density is about 10^{12} Am^{-2}. Power is in tens of megawatts (for an extremely short time). Temperatures in the 10^6 °C range are produced when powers of about 10^{11} W are released into wires of 1 to 75 microns diameter[5].

There are other applications besides tube interiors; e.g, an excellent bond is obtained between two silica rod ends by using a 2500 J capacitor bank at 8 kV to explode a 25 micron thick Ta foil pressed between the rod ends[5], Cu and Si can be welded together similarly. The temperature rise in materials thus welded is negligible.

2 THEORY

An outline only will be presented here. Detailed theory is available in a series of conference proceedings[6-9]. A classification of exploding wires[10] is based on the rate of energy delivery to the wire:

1. Melting: The available energy < that required for complete vaporisation:

$$\frac{1}{2}CV^2 < W + \int I^2 R . dt \qquad \{1\}$$

where C = capacitance of storage capacitor,
V = original potential on it,
W = energy to vaporise wire,
I = current through wire,
R = effective resistance of discharge circuit.
The wire here never vaporises but just breaks up into droplets (fuse behaviour).

2. Slow Explosion: The time needed to vaporise the wire, $t_v \gg$ the time needed for instabilities to develop in the melted wire.: $t_v \gg t'$,
where t_v = time to vaporise the wire, t' time constant of instabilities (= time for instabilities to double in size). Instabilities are, e.g., wire curling effects due to thermal expansion before melting and 'unduloids' (surface tension of molten wires creating connected droplets – but see later)[11]. This means the explosion is slow enough for physical distortions to affect the way it develops.

3. Fast Explosion: The time to vaporise, $t_v \ll$ the time constant of instabilities: $t_v \ll t'$. This means explosion occurs so fast that no significant wire shape changes occur.

4. Explosive Ablation: The time to vaporise $t_v \ll$ the electrothermal time constant t'': $t_v \ll t''$.
The electrothermal time constant is the time for the boiling point temperature to penetrate to a depth of r/e, where r = original wire radius and $e = 2.303$ (natural log base). The wire vaporises in a thin surface film before its centre gets heated. This is due to an effect called the pinch effect (see later) rather than to the radio frequency "skin effect" caused by the high rise time of the current pulse (DC at switch-on is effectively like a single cycle of RF, with ringing at the circuit resonant frequency), well-known in induction heating. The latter may occur on wire rupture however, and it is advantageous to keep the wire radius << the skin depth (see later).
Another classification scheme[12] gives the rate of energy delivery as a factor.
There is a linear relation between wire cross sectional area and time to explode it, at constant initial voltage.

3 PRACTICAL ASPECTS

Typical equipment could be a 20 kV supply and 30 microfarad low inductance capacitors (circuit inductance 0.3 microhenrys). The explosion can be seen with high speed photography (Figure1)[6].
Webb et al.[13] give a good account of the event sequence of how the explosion moves from

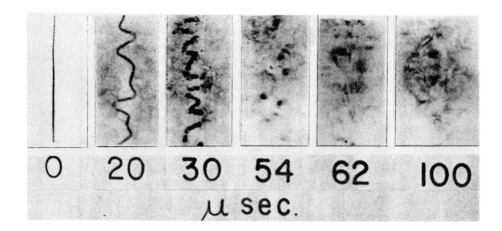

Figure 1 *X-ray of exploding wire phenomenon, 1.0 mm Cu*

outer periphery to inner core of the wire. If the wire radius is small compared with the heated skin thickness effect, it will heat uniformly throughout. Thermal oscillations of the lattice effectively increases the internal pressure in the wire. This would be relieved by thermal expansion when a wire is heated slowly, but for ultra-fast heating the rise in pressure exceeds the maximum expansion rate (which is the speed of sound through the lattice). This causes a very high thermal pressure (10 bar) in the lattice, resulting in an explosion.

Connections to exploding wires may be from a few inches to several feet long, The effect of this is discussed by Maninger[14].

There is an optimum wire diameter to obtain the highest explosion shock wave energy[15], e.g. a 500 J/cm energy input into an 18 mil (18x25 micron) wire gives the maximum energy output, Varying the wire diameter by a few mils either side, decreases the shock energy by > a factor of 5. A 15 mil wire is the optimum for 282 J/cm. Increasing the stored electrical energy has little effect. The existence of an optimum diameter is because very thin wires explode in short times, using little of the available stored energy, whereas thick wires will not absorb enough energy to vaporise them. The latter occurs even if there is enough stored energy to completely vaporise the wire[16].

4 EXPLODING FOILS

Foils may be exploded instead of wires[17,20], e.g. thin metallic films on the inside of glass cylinders can be exploded[20]. This could be applied to coating the exteriors of tubes. Alternatively, tube exteriors could be coated using wires while rotating the tubes rapidly.

Foils typically explode simultaneously all over their surface within about 10^{-7} seconds[18]. This effect can be used to accelerate a thin projectile plate onto an adjacent surface and 80 kbar pressures have been reported[18]. However, 3000 J are needed for a 2 inch square plate on Al foil. A high power installation would typically have 16000 J in 125 kV capacitors of value 2 microfarads. A typical system inductance is 130 nanohenrys and the ringing frequency with a short circuit load is about 330 kHz, implying a time to peak current of 0.75 microseconds[19].

A large thin perspex projectile plate (called a slapper plate) is held just over the foil to be exploded. The explosion shock wave breaks out cleanly an area equal to the exploding foil and slaps it onto a substrate, The electrical energy in the capacitor bank is efficiently transferred (>30%) into slapper kinetic energy. The exploding foil must be on a perspex (or similar) backing plate for best pressure. The projectile plate is also perspex and its front face is covered with a thin sheet of the clad metal foil to be slapped onto the substrate.

5 GENERAL

Temperatures range from 2×10^5 to 2×10^6 °C[21].

Exploding a wire is a simple method of concentrating energy in a region at a rate far greater than it can be removed by ordinary (non-explosive) processes (conduction, convection, radiation). The result is the creation and explosion of a dense superheated plasma. Such a plasma normally exists stably only deep within a star and so it has a violently non-equilibrium and explosive nature elsewhere. An exploding wire is effectively a very hot metallic gas compressed to the volume of a solid, which thus becomes a plasma jet. The unusually high velocity produced leads to very adherent coatings.

An exploding wire plasma (ions etc.) can only expand perpendicularly to the magnetic lines of force created by the very high current in the wire. This contrasts with an arc, where such transverse motion of the plasma there, is impeded by a magnetic bottle effect due to the arc current[22]. For electric arc metal spraying (and flame spraying), a strong blast of gas is thus needed to accelerate the metal plasma transversely onto a nearby substrate. A better and more adherent spray coating is thus obtained by wire explosion and it is well collimated by this and other effects[22].

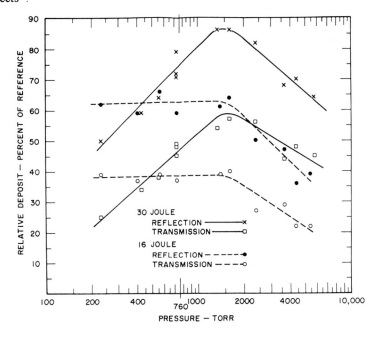

Figure 2 *The curve of deposit with respect to presure*

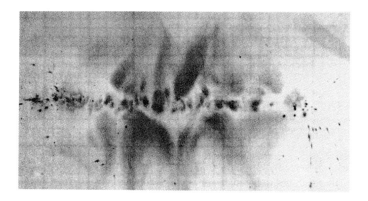

Figure 3 *A high inductance record: 20J, 20μsec, 100μH*

A foil cylinder within a containing tube can only explode inwards (implode) and the effect compresses the initial axial magnetic field of several kilogauss to about 60 kG[23].

"Cladding" is possible for tube bores, if an axial wire is exploded with the tube ends sealed[24]; (the tube ends are left open for ordinary wire explosion coating). High shock pressure will explosively clad a liner onto the tube wall. It is suitable only for thick-wall (strong) tubes (e.g. cylinders for engines); deformation occurs with thin walled tubes.

High speed photography is often used to record wire explosion (Figure 1) but an image can be recorded directly on a paper held within 100 diameters of an exploding 80 micron diameter nichrome wire[25], at pressures from 100 to 5000 torr, with explosion energies from 5 to 50 J (i.e. 5 to 50 times the energy needed to melt the 80 micron 40 mm long wire). White bond paper has suitable porosity, ablative and shock-absorbing properties. Sharp outlines are left on the paper by thermal ablation, and oxide and pure metal condensate deposits. Gravity

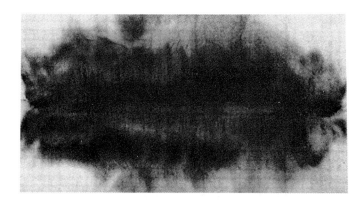

Figure 4 *A low inductance record: 10J, 3μsec, 0.5μH*

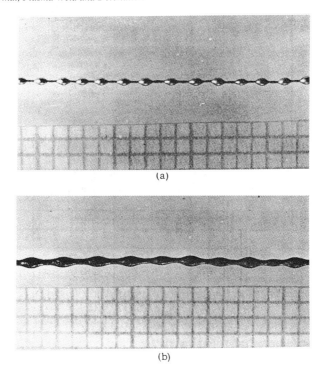

(a)

(b)

Figure 5 *The unduloids with a millimeter grid: (a) 0.32 mm diameter silver, (b) 0.5 mm diameter copper*

has a major effect on the shower of metal droplets, if horizontal apparatus is used, indicating that a vertical wire would be best for coating purposes. The wire snakes as it explodes, with about 5 peaks and 5 troughs per mm (for Cu wire)[26] (Figure 1).

The effect of ambient pressure on the coating deposit is shown in Figure 2 [27]. As expected, lower pressure allows faster diffusion and less energy concentration behind the shock wave,

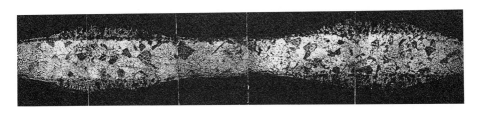

Figure 6 *Longitudinal section of the copper unduloid obtained from the 0.75 mm diameter wire (enlarged 35 ×)*

which gives a more uniform coating, but thinner than at 1 bar. A much bigger effect on uniformity and thickness of coating is system inductance (Cf. Figures 3 and 4)[27]. A high inductance slow explosion is much inferior to a low inductance fast explosion. Figure 4 shows fine detail of the string of molten toroids produced by the exploding wire.

Thin wires burst into unduloids (Figure 5)[28], but thick wires break into segments. It has been suggested[29] that the unduloids are caused by a surface tension effect on the molten wires, but regular hot spots along fine solid W wires heated by an unusual RF method[30] are also observed and these are possibly due to standing current waves caused by a parasitic oscillation (certainly not by surface tension effects). This possibility is supported by other evidence[31]. Wires melt from the outside inwards, not due to the RF skin effect (high frequency wavefront effectively produced at DC switch-on), because the frequency is not high enough, but is due to an effect called pinch pressure[32].

Nasilowski[33] found unduloids after exploding Ag and Cu wires (Figure 5) and deduced that these must have formed after the circuit was broken, because if they had been created while current was flowing, the thin necks between them would have overheated due to their thinner diameter, whereas Figure 6 indicates an even temperature all along the wire. The wire core retains its grain structure, while the unduloids show recrystallisation and dendrites (due to melting and freezing); see Figure 6 [33]. However another explanation of the unduloids may be the above-mentioned regular hot spots mentioned in the previous paragraph, if their lifetime was short.

References

1. T, Suhara, K. Kitajima, S. Fukuda and H. Ito: Proc 7th Int Metal Spraying Conf, London, 1973, paper 26.
2. T. Suhara, S. Fukuda and H. Ito: Proc 6th Int Metal Spraying Conf, Paris, 1970, paper B3.
3. K, Kase, H, Ito and Y. Mihashi: *Jpn Soc Powder Powder Metall*, 1970, **16**, 338.
4. S. Hasui: Yokendo, Tokyo, (1960). p. 150.
5. V. E. Scherrer, in W,G, Chace and H,K. Moore (Eds), 'Exploding Wires', Conf at Boston, (Plenum, New York, 1959). p. 118 .
6. 'Exploding Wires', conference 1959, at Boston, Eds W.G. Chace and H,K, Moore, publ by Plenum Press, New York, 1959.
7. Ibid. Idem, 2. 1962. at Boston, published by Plenum Press, 1962.
8. Ibid, Idem, 1964, at Boston, published by Plenum Press, 1964.
9. Ibid, Idem, 1967, at Boston, published by Plenum Press, 1968.
10. W. G. Chace and Levinej, *J. Appl. Phys.*, 1960, **31**, 1298.
11. J. Nasilowski, Ref. 8, p. 295.
12. F. H. Webb, Ref. 7, p.1.
13. F. H. Webb et al, Ref.7, p. 37.
14. R. C. Maninger, Ref. 7, p. 121.
15. D. L. Jones et al, Ref. 7, p. 140
16. H. S. Leopold, Ref. 8, p. 134.
17. G. J. Woffinden, Ref. 8, p. 194.
18. D. V. Keller and J,R, Penning jr, Ref. 7. p. 263.
19. A. H. Guenther et al. Ref. 7. p. 282.

20.   R. S. Dennen and L.N. Wilson, Ref. 7, p. 155.
21.   O. H. Zince et al, Ref. 8, p. 103.
22.   J. Rothstein, Ref. 8, p. 120.
23.   G. Schenk and J,G. Linhart, Ref. 8, p. 223.
24.   I. M. Fyfe and R. R. Ensminger, Ref. 8, p. 262.
25.   K. E. Moran, Ref. 8, p. 285.
26.   W. G. Chace et al, Ref. 6, p. 59.
27.   K .E. Moran, Ref. 8, p. 292.
28.   J. Nasilowski, Ref. 8. p, 29.
29.   H. W. Baxter, 'Electric Fuses', publ, by E. Arnold, London, 1950.
30.   P. S. Sidky and M.G. Hocking,, 'Contactless Heating of Thin Filaments',World Patent W089/1177,AU No.615314, TWD No.43909, ES No. 2011743, NZ No. 229175. etc.
31.   F. D. Bennet et al, Ref. 8, p. 71.
32.   E. B. Carne, *Trans AIEE*, 1953, **727**, 593.
33.   J. Nasilowski, Ref. 8, p. 305.

Section 2.3 Laser and EB

2.3.1
(Nickel Base Alloy + Cast Tungsten Carbide Particles) Composite Coating by Laser Cladding and the Wear Behaviour

Wang Peng-zhu[1], Ma Xiong-dong[1], Ding Gang[1], Qu Jing-xin[1], Shao He-sheng[1] and Yang Yuan-sheng[2]

[1]BEIJING GRADUATE SCHOOL, CHINA UNIVERSITY OF MINING & TECHNOLOGY, CHINA

[2]METAL RESEARCH INSTITUTE, CHINESE ACADEMY OF SCIENCES, CHINA

1 INTRODUCTION

The laser has brought great changes in many fields of our life since its invention. Lasers can project highly integrated "pure" energy, according to the required amount, position and time, to any targeted material exactly. In surface engineering this has produced laser (strengthening) technology, including laser transformation hardening (laser quenching), laser melting, laser surface alloying (laser chemical heat treatment), laser coating (laser cladding), laser induced chemical/physical vapour deposition (LCAD/LPVD), etc. Laser surface technology has been developed very quickly[1-5], as it possesses incomparable advantages: super heating speed, super high temperature; small distortion of the work piece since the heat is integrated into a small area; fine micro structure of laser quenched part and high hardness than that by conventional heat treatment Using lasers, super high temperature ceramics or metal-ceramics composites could be coated onto plain steel to obtain wear corrosion and heat resistant properties. Laser processing can be controlled by computer easily to realize automatic production and laser processing is a "green project", which does not pollute the environment.

Wear is the number one failure mode of many mining machine parts. Considering the special working conditions and failure mechanisms, laser coating technology is regarded as being more suitable for surface protection of mining machine parts.

Laser cladding of nickel based, self fusing alloys has been successfully applied to balance plates of slurry pumps[6]. In order to increase the service life further, attention has been focussed on coatings made up of ceramics, which have higher wear, corrosion, and heat resistance than other materials. But, it seems too difficult to produce applicable ceramic coating by laser cladding due to the very special physical, especially thermal physical properties of ceramics. Thus composite coatings of many, metal-ceramics systems have been developed and used, instead of pure ceramic coatings. The present paper deals with the manufacture, structure analyses and wear behaviour investigation of the composite coating made up of nickel based alloy (NBA for short) and cast tungsten carbide (C-WC for short) particles.

2 EXPERIMENTAL CONDITIONS AND PROCEDURES

2.1 Experimental Materials

AISI 1045 steel was used as substrate materials[1], the size of the specimen was 10 × 10 × 100 mm. The nominal compositions of the steel, the nickel based alloy and tungsten carbide are shown in Table 1.

2.2 Coating Manufacturing

2.2.1 Coating Preparation. A certain amount and size of NBA and C-WC powder were mixed with a proper amount of adhesive slurry, which was produced by 2% sodium carboxmethlcellulose and 98% water, to form a past. The paste was spread onto a 1045 steel specimen, and then oven-dried at 120°C for 20 minutes. The thickness of the paste was 1.2mm.

2.2.2 Laser Cladding. Laser cladding, was performed by use of a CW 3kW CO_2 Laser. The diameter of the beam was 4 mm, the scanning patterns included single line and overlap, the overlap ratio was 20%.

2.2.3 Structure Analyses and Hardness Test. D/max-RB type X-ray, diffractometer, S-250MK3 type SEM were used to analyse the phase components and microstructures. EDAX-900 type EDS was used to analyse the micro distribution of the composition of the coating. Micro Hardness tester and Rockwell hardness testers were used to measure hardness.

2.2.4 Wear Tests. A sliding wear test was performed on an MM-200 wear tester. The laser clad specimen was fixed, the rotating ring made of quenched Gr15 steel was pressed onto the coating with load L = 300N, the relative sliding speeds were 0.42m/s and 0.84m/s. The sliding contact surfaces were grounded to Ra = 0.8 μm. No lubrication was applied.

An erosive abrasion experiment was carried out with an MLS–23 type rubber wheel wear tester. Slurry composed of water, coal and quartz sand was used as a wear medium in order to simulate working conditions of mining machine parts. The load was 70N, and the relative speed was 1.67 m/s. In both cases, the wear resistance was measured by worn volume over wearing time (per hour), Vw (mm^3/h).

Table 1 *The composition of experimental materials*

	C	B	Cr	Si	Mn	Fe	Ni	W	Melting point
1045 Steel	0.47	-	-	0.31	0.62	bal.	-	-	1450–1500°C
NBA	0.84	3.62	1.57	4.33	-	4.27	bal.	-	1084°C
C–WC	4.2	-	-	-	-	-	-	95.8	2525°C

The size of NBA particles is 40–100 μm. The size of C–WC particles is 50, 10, 200, 300, 400 μm, the content is 30%, 60% wt.

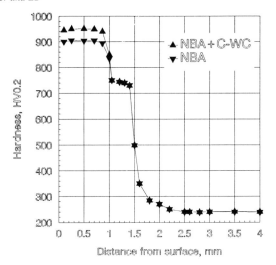

Figure 1 *Hardness distributions of NBA and NBA+ C-WC (50μm size) coatings*

3 LASER PROCESSING EXPERIMENTS

3.1 Experimental Purpose

The purpose of the experiments was to obtain a NBA + C-WC composite material coating. The tungsten carbide particles would maintain their original structure, that is to say, C-WC particles could not be dissolved into the NBA matrix when heated by laser beam. According to wear theory, the (strong matrix + hard second phase particles) composite coating, plus metallurgical (not mechanical) adhesion with the substrate, was expected to show the highest wear resistance.

3.2 Experiments with C-WC Particles of Different Sizes

The laser cladding conditions were kept as: output power P = 1200 W, diameter of laser spot D = 4 mm, scanning speed V_s = 3 mm/s. The results are shown in Table 2.
The experimental results showed under certain laser processing conditions, there existed a

Table 2 *The laser processing results with C-WC particles of different sizes*

Size (μm)	Results
50	Totally dissolved into the matrix, original particles could nt be observed by metallography
100	Most dissolved, a few of the original particles observed on the cross section, they stayed along the edge of the scanning strip (Figure 2)
200	Not dissolved, except the tips of the particles
300	Not dissolved, except the tips of the particles

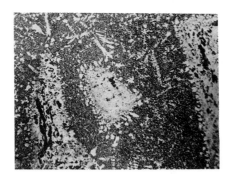

Figure 2 *The dissolution of a C-WC particle with the size of 100μm (× 100)*

minimum size for C-WC particles to form composite coating with NBA matrix. In this paper that is 100μm. Metallographical examination by microscopy and SEM showed there was little difference between the micro structures of NBA coating and (70%NBA + 30%C-WC) coating when the size of C-WC was 50μm, but the hardnesses were different, as shown in Figure 1. These results form the solution hardening of W and C atoms. The authors of this paper take it that in this case the function of tungsten carbide has composition adjustment, not composite formation. Figure 2 shows the presence of a C-WC particle after laser cladding, with the size of 100μm.

3.3 Experiments with Different Input Power Rate

Theoretically speaking, increasing the scanning speed will make it possible for smaller C-WC particles to compose composite coating, but the increase of the speed was limited to a minimum value, at which the matrix (nickel base alloy) was expected to melt and adhere to the substrate by welding (metallurgical bounding). The decrease of scanning speed will also

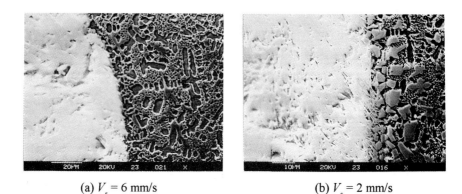

(a) V_s = 6 mm/s   (b) V_s = 2 mm/s

Figure 3 *The scanning speed (V_s) influences the dissolution of C-WC particles; (a) and (b) show the dissolution tendency of C-WC particles when input power increases, the sizes of C-WC particles were the same*

influence the composition and micro structure of the coating: when the scanning speed decreases, the temperature of melting pool will increase. On the one hand the diffusion of element across the boundary will be accelerated, therefore more iron atoms will penetrate into the coating, which will affect the wear as well as corrosion property; on the other hand the viscosity will decrease and this will make the C-WC particles sink and gather on the bottom.

It should be possible to establish an quantitative relation between the size of C-WC particles and the proper laser processing parameters suitable for forming composite. To make the analyses simple, power input rate e is defined as the laser power absorbed by the coating per unit area, $1 m^2$, during a single line scanning, that is:

$$e = \frac{4P}{\pi d^2} \cdot \frac{d}{V_s}$$

i.e.
$$e = \frac{4P}{dV_s}$$
{1}

here, e = energy absorbancy (J/mm^2), P = output power of laser (kW), d = diameter of laser beam (mm), V_s = scanning speed (mm/s).

In theory the minimum diameter of C-WC particle for composite coating ϕ_{min} is a function of e, $\phi_{min} = f(e)$, but, in fact ϕ_{min} is always decided by experiment.

4 STRUCTURE ANALYSIS

X-ray diffraction revealed the phase constituents of NBA + 60% C-WC coating. The coating was composed of WC, W_2C, γ-Ni, Cr_7C_3 and $Cr_{23}C_6$.

Table 3 shows the EDAX experimental result, revealing the composition change across the border between C-WC particles and NBA matrix, the specimen is shown in Figure 3(a).

5 WEAR EXPERIMENTS

5.1 Dry Sliding Wear Test

Table 4 is the wear experimental results on MM-200 wear tester. The wearing rate of composite

Table 3 *The element distribution across the border of C-WC and NBA.*

	On C–WC particle		On border ($\phi 1 \times 1\mu m$)		In matrix	
	wt%	at%	wt%	at%	wt%	at%
W–M	99.01	93.75	97.90	93.42	18.59	6.24
Ni–K	0.37	1.11	0.40	1.20	53.68	56.39
Cr–K	0.15	0.51	0.19	0.65	12.44	14.76
Fe–K	1.47	4.63	1.50	4.72	10.06	11.11
SI–K	-	-	-	-	5.24	11.5

Figure 4 *Worn surface of NBA + fine C-WC coating (× 200)*

coatings are much (average value 10.7 times) lower than that of nickel based alloy. In order to compare the influence of particle size, two groups C-WC particles were used. It can be seen that when relative sliding speed was 200 rpm the wearing rates of the two kinds coating were the same, but when the speed was 400 rpm the coating with coarse C-WC showed higher wear resistance. The mechanism was that at high speed the temperature of sliding surface increased, a layer of NBA matrix of certain depth softened. When the depth of the layer reached the diameter of C-WC particles, the particles were removed and stripped off. This will not only lose the reinforcing function of the ceramics, but also aggravate the wearing of materials, since the separated ceramic particles become hard abrasive. Figure 4 shows the typical worn surface.

5.2 Erosive Abrasion Test Results

Figure 5 shows that with the increase of WC-C particle's content, the erosive abrasion resistance of the coating increases.

6 CONCLUSIONS

1. To form metal-ceramic composite coating by laser cladding, the sizes of the particles and the laser processing parameters are the key factors when the matrix metal is decided.

Table 4 *Dry sliding wear results, volume wear rate Vw (mm³/m)*

Load L (N)	300	300
Speed Ur (rpm)	200	400
NBA coating	0.55	1.3
NBA + coarse C-WC (size 300 μm) -	0.050	0.075
NBA + fine C-WC (size 150 μm)	0.050	0.14

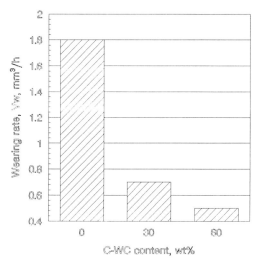

Figure 5 *Erosive abrasion test results*

2. Under certain laser processing conditions, there exists a minimum size for tungsten carbide particles to form composite coatings with metals.
3. Dry sliding wear tests show that the size of the ceramic particles influences the wear resistance of the composite coating, coarse particles have stronger reinforcing effects.
4 Slurry abrasion tests show that the wear resistance increases with the content of tungsten carbide particles.

References

1. W. Cerri et al., *Surface and Coating Technology*, 1991, **49**, 40.
2. N. Ax én et al., *Wear*, 1992, **157**, 189.
3. Chande T, Maztunder, *J. Appl. Phys.*, 1985, **57**, 6, 2226.
4. YANG Xichen, *Acta Metall. Sin.*, 1992, **28**, 2, B94 (in Chinese).
5. A.F.A. Hoadley et al., *Metallurgical Transactions B*, 1992, **23B**, 631.
6 WANG Pengzhu, Thesis for Doctor Degree, Apr. 1994 (ln Chinese).

2.3.2
Electron Beam Surface Hardening of 42CrMo4 Steel

T. Jokinen and I. Meuronen

VALMET AVIATION INDUSTRIES/ENGINE MAINTENANCE, P.O. BOX 10, 37241 LINNAVUORI, FINLAND

1 INTRODUCTION

A hard, wear resistant surface combined with a tough base material is required to provide useful life for engineering components. This kind of combination of features can be reached with surface hardening methods. Electron beam (EB) surface treatments are included in the most promising surface technologies[1]. Commonly used EB surface treatments are hardening, melting, alloying and cladding. The heat of the processes is generated by high velocity electrons hitting the work piece and then changing their kinetic energy to heat the work pieces. The charged electron beam is deflected very rapidly electromagnetically on the work piece's surface and so with the moving work piece the treatment area becomes larger. In EB hardening the surface is keeping its solid state while in the others the surface is melting more or less. As a consequence of this the machining after EB hardening is not normally needed.[2, 3]

In electron beam hardening both heating and cooling are very rapid. Fast heating is a result of good absorption of the EB; approximately 75 % of the power generated is converted to heat when the beam is incident perpendicular to a steel surface. Energy transfer to the interior of the material is affected by conduction, then rapid cooling of the austenite required for martensite formation occurs by self-quenching. Depending on the material selected the work piece thickness required should be 5 to 10 times the austenitizing depth[2].

Typical hardening depths reached by electron beam are 0.1–1.5mm depending on the parameters and material. Suitable materials for EB hardening are martensitically hardenable ferrous materials[2].

Due to precise and sharp heat source the electron beam hardening is very suitable to local treatment of complex components. Costs savings in EB hardening are consisting of unnecessary machining or treatments after the hardening; extremely low hardening distortions, clean and bright surface, small changes in surface roughness are characteristics for the EB hardening. Also tempering is not usually required after EB hardening; however it is necessary for some high alloyed steel grades[1].

Wear resistance of engineering components has been increased by electron beam hardening. This is a result of increased hardness and decreased grain size.[1, 2, 3]

The aim of the study[4], which was the basis of this paper, was to find proper parameters of electron beam machine for using EB hardening with 42CrMo4 steel.

Table 1 *Chemical composition of 42CrMo4 steel used in tests*

	C	Si	Mn	P	S	Cr	Mo	Cu	Al
%	0.43	0.22	0.79	0.02	0.015	1.07	0.25	0.14	0.03

2 EXPERIMENTAL

2.1 Material and Test pieces

The test material to be hardened was 42CrMo4 steel, which is medium carbon Q&T steel. This steel is very commonly used on transmission components for example in axles, gears etc. where though and strong base material is needed while the surfaces have to be hard and wear resistant. The analysis of the test material is shown in the Table 1.

Test pieces (Figure 1) were quenched and tempered to hardness 310–320 HB. One test piece was in soft annealed condition. After the heat treatment the surfaces of the test pieces were machined so that the layer of eventual decarburation was removed. The surfaces to be hardened were ground so that the changes in surface roughness were measurable. For each test piece accurate holes for measuring the dimensions before and after the hardening process for distortions were made. The tests pieces were fastened so that the distortions could happen almost freely.

2.2 Electron Beam Machine

The electron beam welding machine used in tests is made and installed by Messer Griesheim (nowadays IGM Roboter Systeme) in 1992. The machine and its main technical information, is shown in Figure 2.

Constant parameters used in tests were:
- acceleration voltage 100 kV
- working distance 500 mm
- vacuum level 1×10^{-2} mbar
- focusing current 1858 mA

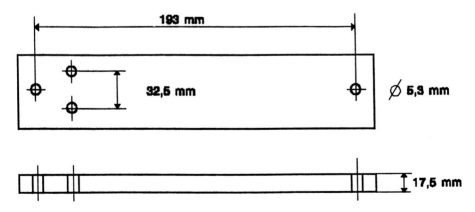

Figure 1 *Test piece dimensions*

Technical Data

Beam power:	max. 30 kW (150 kV, 200 mA)
Working chamber:	11 m³ (2700 x 2000 x 2100 m³)
Vacuum equipment:	3 mechanical pump, cryogenic pump
	10^{-4} mbar
Workpiece manipulators:	XY-table, Z-table, Rotary device,
	Tilting device
Specialities:	Filler wire feed unit
	Seam tracking control unit
	Pulsed & modulated beam
	Horizontal position for EB -gun
	EB-gun column movement

Figure 2 *Electron beam machine used in tests*

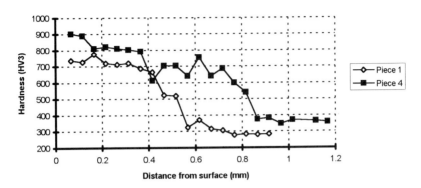

Figure 3 *Hardness profiles of test pieces 1 and 4. Test piece 4 was quenched and tempered before EB hardening, testpiece 1 was left in soft annealed condition. Parameters: I = 15 mA, v = 7 mm/s*

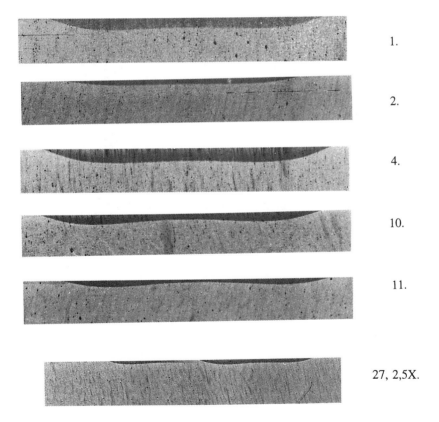

Figure 4 *Macrographs of hardened layer in test pieces. Nital 3% 5.6 x*

- deflection; ellipse 13×3 mm, frequency 1000 Hz

Variable parameters; beam current and velocity, used in tests were varied between 10–50 mA and 4–85 mm/s (36–460 cm²/min).

2.3 Examinations

After EB hardening test pieces were cut, ground, polished and etched by Nital 3%. Both macro- and microphotographs were taken and the hardness profiles were measured by Vickers method using load of 300g. Hardening depth was determined as the distance from the surface to depth, which hardness is above 500 HV03. Because the wear resistance is very comparable to the hardness of the surface, it was estimated upon hardness value.

Changes in the dimensions were measured by 3-D measuring machine, which volumetric accuracy is 0.006mm and accuracy by axle is 0.0025mm. Measured magnitudes were: shrinkage of length and width, straightness and flatness.

Surface roughness was measured also before and after the hardening test. Measurements were made by digital surface roughness meter. Measured parameter was average roughness, R_a, from the length of 15mm crossing the hardened track.

Microstructure of the hardened layers and base material was determined by optical microscope and hardness results.

3 RESULTS

3.1 Hardness Values and Profiles

3.1.1 Soft Annealed Test Piece. According to Figure 3 the hardness of the hardened layer is little over 700 HVO₃, which is three times harder than the base material. Hardening depth is about 0.5 mm. Length, in which hardness decreases to the level of the base material, is about 0.2 mm. Width of the hardened layer is, according to Figure 4, 17mm. The bottom of the hardened track is quite flat.

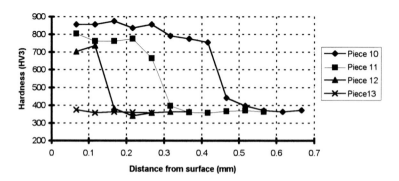

Figure 5 *Hardness profiles of test pieces 10, 11, 12 and 13. Parameters: I = 20 mA.*
10) v = 14 mm/s, 11) v = 16 mm/s, 12) v = 19 mm/s, 13) v = 23 mm/s

3.1.2 Quenched and Tempered Test Pieces. Higher hardness values were achieved by hardening the tempered test pieces than soft annealed. This is a result of carbon distribution and its physical state. In tempered condition the carbon is more uniformly distributed than in the soft annealed condition, so diffusion distance of carbon to austenite is shorter. In the soft annealed condition the carbon is in more stable state than in tempered one, so it dissolves quite slowly to austenite. On the contrary in the tempered condition the carbon is quite unstable so it is quite quick to dissolve. This can be seen clearly in the Figure 3; hardening depth is deeper with test piece 4 than test piece 1, also hardness values are higher, although parameters are the same. So in both test pieces the temperature in every distance to the surface has been the same, it is just the state of the carbon, which orders in this case the hardness and the hardening depth. The same temperature distribution can be seen in the macro-structure of the hardened layer (Figure 4); both test pieces have the same contour of the layer.

Hardening depth depends greatly on the heat input. In Figure 5 is shown hardness profiles of test pieces, in which the beam current has been the same and travel speed has been changed. A lower speed produces deeper hardening. Similarly, with beam current with the same speed. With more heat to the surface, the deeper hardening. In both cases the limits are non-hardening and surface melting.

In a few test pieces the contour of the bottom of the hardened track is not uniformly flat, (Figure 4) as a consequence of the wrong heat input to the track. In these samples too little heat is converted to the middle of the track. This depends on the oscillation amplitude, contour, frequency and travel speed. Normally more heat has to come to the edges of the track, because the conduction of the heat is three dimensional on the edges and in the middle of the track it is two dimensional.

If there is a need for larger areas to be treated, it is possible to use overlapping EB hardened tracks. Although the overlapping distance and other parameters are accurately estimated or tested there is always an annealed area between the two tracks, as in Figure 6. In Figure 6 is also shown that the hardening depth has increased in the second hardening tracks. This is

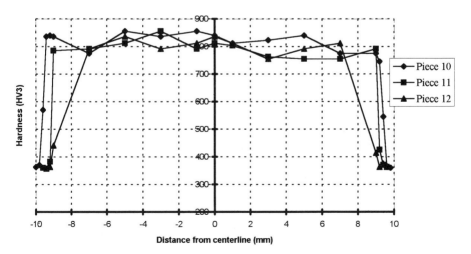

Figure 6 *Cross-sectional hardness profile of test pieces 10, 11 and 12. Parameters: I = 20 mA. 10) v = 14 mm/s, 11) v = 16 mm/s, 12) v = 19 mm/s*

because the work piece has been warmer than in the beginning but still there has been enough cold material for self-quenching. Annealed area between two tracks is not necessarily a disadvantage; the softer area will wear first and then in lubricated conditions it could be like an oil pocket, which decreases wearing.

Figure 7 shows a cross-sectional hardness profile; the hardness is very steady in the hardened layer, as in longitudinal direction.

3.2 Microstructure

3.2.1 Soft Annealed Test Piece. In the soft annealed condition the microstructure of the test material is pearlitic/ferritic according to hardness value (230 HV_{30}) and microscope inspection. In the hardened layer the microstructure is bainitic/martensitic. (Figure 8) The microstructure of the hardened layer is non-uniform resulting from the original phases and a very rapid heating of the process.

Grain size in the base material was 20–52 μm, while in the hardened layer it was 20–30μm. Because of very rapid heating grain growth of austenite is eliminated and with very fast quenching fine grain size remains. Small grain size is very good for wear resistance, because it gives more thoroughness to the hardened layer without losing its hardness.

3.2.2 Quenched and Tempered Test Pieces. In these test pieces the base material's microstructure was tempered martensite and in the hardened layer it was un-tempered martensite. As a consequence of preliminary microstructure the hardened layer is total martensite, which hardness is very high.

In the hardened layer the grain size was 13–19μm, while in the base material it was 3–20μm as a result of rapid quenching.

3.3 Surface Roughness

After the process the surface roughness was a little worn compared to the situation before treatment. The difference is shown in Table 2. Average increases in R_a-values were 15 % in cross direction and 20 % in longitudinal direction to the hardened track. It was seen quite

Figure 7 *Microstructure of the hardened layer in test piece 1. Nital 3%, 500 x*

Figure 8 *Microstructure of the hardened layer in testpiece 2. Nital 3%, 500 x*

clearly that the surface roughness increased with heat input.

For each test piece the surface roughness was almost the same as a consequence of martensite reaction; martensite has a different volume compared to the other phases.

Achieved values of surface roughness (R_a < 1 μm) are compared to careful turning or not so careful grinding. Applications to this roughness are for example bearing bush (longitudinal movement), coupling- and brake surfaces, gears, torsion axles.

In overlapped tracks the surface roughness values were a little rougher than in one track, because in measurement the overlapped areas were taken in to the measuring area. Noticeable is also the roughness before tests.

Because the EB hardening procedure was done in vacuum the surfaces of the work pieces are normally clean and no oxidation or contamination has happened, although the hardened

Table 2 *Surface roughness before and after EB hardening. Test pieces 27 and 28 contain overlapped hardening tracks*

Number	Cross-sectional [R_a]		Longitudinal [R_a]	
	Before	After	Before	After
1	0.792	0.993	0.574	0.74
2	0.802	0.95	0.512	0.764
3	0.8	0.83	0.472	0.506
4	0.81	0.913	0.442	0.7
10	0.782	0.998	0.488	0.668
12	0.878	0.915	0.59	0.664
13	0.834	0.860	0.538	0.572
27	1.102	1.245	0.708	0.780
28	1.3	1.463	0.74	0.928

tracks were clearly seen by human eyes. No hardening cracks were seen visually or by microscope.

3.4 Distortions

Measured changes in length, width, straightness and flatness are shown in Table 3. Generally in these tests the distortions of the work pieces were quite small. This shows the optimal and sharp heat input of EB hardening. The biggest changes were measured in soft annealed test piece as a consequence of the lower strength of the soft annealed test piece compared to the tempered ones. Naturally in overlapped test pieces the reason for greater distortions were higher heat input.

Distortions are a result of volumetric changes occurring during austenization and martensite reaction. Also the difference in temperatures causes distortions. Normally distortions are comparable to amount of heat transferred to work piece. But several phenomena affect the amount of distortions, for example strength, dimensions and shape of the work piece. So it is very difficult to give general estimation programme to the distortions.

4 DISCUSSION

The wear resistance of a material depends greatly on the wear mechanism occurring. According to this it is very difficult to know a material's wear resistance exactly without very careful and real conditions simulating tests. Still it is possible to estimate wear resistance from hardness values. When the background is the hardness, the test pieces in this experiment are very wear resistant. For all test pieces the hardness of the surface increased a minimum two times, in some cases over three times harder than the base material. The highest hardness values were about 900 HV. Material hardness after quenching depends mainly on carbon content of austenite. So although EB hardening is a very rapid process there is still sufficient time for austenite to dissolve. Also very fast cooling and high compressive stresses increase hardness values. The hardness values are similar in every test piece so this is the specific hardness to this material in EB hardening, depending on the condition of carbon and base material.

Table 3 *Measured changes in length, width, straigthness and flatness. Test pieces 27 and 28 contain overlapped hardening tracks*

Number	Shrinkage of length (μm)	Shrinkage of width (μm)	Straightness (μm)	Flatness (μm)
1	46	83	138	143
2	11	41	80	86
3	1	10	55	58
4	32	49	86	90
10	20	41	95	101
12	13	50	93	96
13	17	40	77	82
27	22	54	132	126
28	13	73	111	101

Very fine microstructure gives to the surface better toughness without losing the hardness, so it raises wear resistance. A very fine microstructure in the EB hardened layer is the result of very rapid heating to above of the austenite temperature, a very short dwell time and very rapid cooling. In this kind of fast process it is impossible for austenite grain growth. Tempered base material is very suitable for EB hardening; formation of austenite is rather fast, carbon can easily dissolve to austenite and the base material has enough strength to carry wear loads, although hardness values are quite similar to the annealed test material.

Hardening depth depends on the heat input. In every region where the temperature has reached the austenization temperature and has been long enough there, the final microstructure contains hard phases. The limiting factor in heat input is the melting of the surface, although by surface melting it is possible to achieve deeper hardening depths, because after melting machining is needed a reasonable surface finish.

The profile of the hardened layer depends on the material's heat conductivity and the means of heat input. The deflection procedure used in the tests (ellipse) usually gives more heat to the edges of the hardened track.

When a really high toughness is needed in the hardened layer for wear behaviour, or resistance against impacts is needed, then the hardened track is recommended to be tempered. The tempering can be done by the electron beam machine in a similar manner as the hardening was done. Also in some cases the difference between hardness values in the hardened track and the base material is so great that there is a point of discontinuity concerning the fatigue – here tempering is also recommended.

As a consequence of the very little changes in dimensions and the surface roughness, normally the components are ready to use immediately after the electron beam surface hardening. If more accuracy is needed in surface finish, then a careful optimization is needed between hardening depth and process properties.

5 CONCLUSIONS

EB is a very suitable method for surface hardening of 42CrMo4 steel, although careful optimization should be done with process parameters and the hardened layer. The average of 0.4 mm hardening depths were achieved, while hardness in this area was up to 900 HV_3. No significant difference in hardness results was seen between annealed and tempered test pieces, although the tempered one is better for use in wear conditions because of its strength. Grain size in hardened layer was very fine, 13–19μm in tempered test pieces. Small distortions and changes in surface roughness allow components to be normally ready for use after electron beam hardening.

References

1. R. Zenker, 'Electron Beam Surface Treatment. Industrial Application and Prospects', Report of 5th International Conference on Welding and Melting by Electron and Laser beams, 1993, La Baule, France.
2. S. Schiller, S. Panzer, B. Furcheim, 'Electron Beam Surface Hardening', ASM Handbook, Vol. 4, Heat Treating, ASM International 1991, First Printing.
3. J. D. Ferrario, W. J. Farrel, 'A Computer-Controlled, Wide Band Width Deflection System For Electron Beam Welding And Heat Treating', Report of 4th International

Colloqium on Welding and Melting by Electron and Laser Beams, 26-30 September 1988 Cannes France.

4.   T. Jokinen, 'Improving the Wear Resistance of Machine Components with Electron Beam Surface Treatments', Lappeenranta University of Technology, Department of Welding Technology, Master of Science Thesis, 1995.

Section 2.4 Peening, Solar and Other

2.4.1
Controlled Shot Peening. Scope of Application and Assessment of Benefits

G.J. Hammersley

METAL IMPROVEMENT COMPANY INC., UK

1 INTRODUCTION

Shot peening is a cold working process in which the surface of the component is bombarded with spherical shots. Each piece of shot to strike the surface acts as a tiny peening hammer, imparting to the surface a small indentation or dimple, yielding the material and inducing a residual compressive stress. The magnitude of the compressive stress can be as high as 60% of the Ultimate Tensile Strength of the material being treated. Since most fatigue failures initiate from some features on the surface the very considerable degree of prestress afforded by shot peening has a profound improvement on the fatigue limit of the component.

Cold working can also have beneficial effects due to work-hardening, improving grain structure, improving surface texture, closing porosity and testing the bond of coatings. Hence the benefits obtained from the process, in addition to improved resistance to mechanical fatigue, include resistance to stress corrosion cracking, intergranular corrosion, hydrogen embrittlement, fretting and galling, as well as other well proven but somewhat novel applications.

2 HISTORY AND DEVELOPMENT

Metal workers have known intuitively for hundreds of years that cold beating of metals had benefits in strength and hardness, evidenced by the forging of swords. But it was not until a chance discovery in the 1940's that modern shot peening commenced in earnest.

J. O. Almen, working to enhance the fatigue life of car engine valve springs found by accident that shot-blasting produced a significant improvement in fatigue strength. He went on to discover that if he substituted a random shot-mix of grit-like particles with spherical media, all of the same size and properties, then the improvement was even more noticeable. Almen further developed the process to devise a calibration system to ensure repeatability of processing. This method, which became known as the "Almen Strip System", is largely unchanged today, and brings together the variables of the process,viz.,shot hardness, size, density and velocity into a single, measurable quantity, known as "Almen Intensity". This involves the use of spring steel test strips, which are peened in representation of the component to be treated[1]. Since the underside surface of the strip is masked by the rigid steel block on which it is mounted, and the upper surface is cold worked, then the residual stress so induced manifests itself as a curvature in the strip when removed from the test block (Figure 1). The

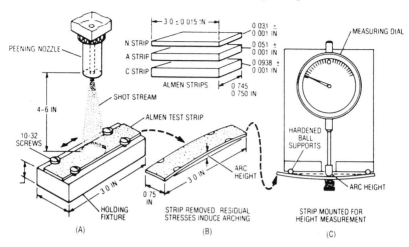

Figure 1 *The Almen strip system*

peening of subsequent strips enables the relationship between the arc-height of curvature and exposure time to peening to be plotted. To begin with the relationship is linear, but ultimately there is a pronounced "knee" in the curve, beyond which very little further deflection on prolonged exposure to peening is noted. This is termed the *"Saturation point"* and for standardisation purposes is usually defined as the first point at which a further doubling of the time of exposure to peening increases the arc height by less than ten percent (Figure 2). Achievement of the desired "Almen Intensity", which is the arc height at saturation point is the most important of the three parameters that define a shot peening specification.

The choice of a suitable type and size of shot is also an important consideration. The range of shots available is usually referred to collectively as "media". Peening shot comes in a variety of types, the most common being in cast steel. The size designation is commonly defined in tenths of a thousandth of an inch, hence "230 shot" has a mean diameter of 0.023

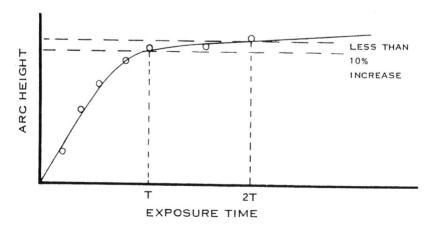

Figure 2 *Saturation curve*

inches. Cast steel shots are available from 0.007 to 0.25 inches and have designated hardness grades of "regular" (45 to 55 HRC) and "hard" (55 to 65 HRC). Media is also available as glass beads which range in size from 0.002 to 0.033 inches. Stainless steel and ceramic media are also available, Having determined the shot size, type and intensity it is necessary to specify the "coverage" – this is a measure of the degree of cold work applied to the surface, and it is generally important, for fatigue improvement, to ensure that the dimples on the surface overlap each other to give a uniform layer of compressive stress. Once the original surface has been obliterated by overlapping dimples the part is said to have achieved 100% coverage. To increase the work hardening effect, and for other metallurgical reasons, it is often necessary to specify a coverage of greater than 100%, and coverage specifications of up to 400% are common. 400% is achieved in practice by peening the part for four times as long as it takes to achieve complete or 100% coverage.

Once an effective shot peening specification has been determined for a given component, then it is a relatively straightforward matter to ensure repeatability by calibrating the machine using the Almen Strip system. But it is in the determination of the optimum parameters that the the most recent developments have concentrated. There remains no more reliable method than fatigue testing of samples and products to arrive at the best solution, but along with more effective fatigue testing methods, and measurement of residual stress by such techniques as X-ray Diffraction (XRD), we now have tools to arrive at a narrow band of shot peening parameters to scope down the variables and embark on a more manageable test programme. Recent developments have enabled mathematical modelling of residual stress profiles by computer. This technique brings together the mechanical properties of the material to be peened and the shot peening parameters. Close agreement has been achieved between the modelled results and measurements made on many samples by XRD, neutron diffraction and hole-drilling techniques. Modelling of the fatigue strength of components is not yet possible but after 50 years of experience a very good idea of the likely improvement to be made by shot peening is available on a wide range of engineering components.

3 PEENSTRESS

This is the name given to a software model developed jointly between Metal Improvement Company and a French Advanced Engineering School in the late 1980's. It is based on Wohlfahrt's Hertzian pressure theory and Zarka's method which determine residual stress from the elastic stress field given by Hertz's theory. The method considers shot peening as a cyclic loading and yielding of the surface by repeated impacts of the shot at velocities up to 200m/s. The theoretical model was first proposed by Guechichi et al[2] and modified by Khabou et al[3] in 1989.

The model makes the following assumptions:

i. The shot is, and remains spherical throughout contact with the surface.
ii. The shot impinges normal to the surface.
iii. The shot is uniform in size.
iv. The loading is pseudo-cyclic.
v. The peened material is stable and exhibits elasticlplastic behaviour.
vi. No prior stress state exists.

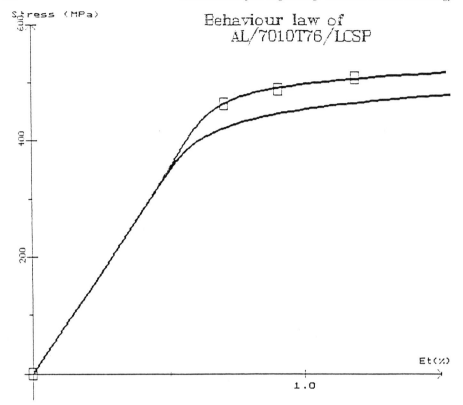

Figure 3 *Stress/strain behaviour law*

The residual stress is calculated in two stages. Firstly the software calculates the shot velocity based on the Almen intensity specified, and secondly, together with the velocity, the shot characteristics and base material behaviour are used to determine the residual stress.

A typical printout is shown in Figure 3.

The PEENSTRESS software contains a library which is a database of materials and their properties. New materials can be added at will, and the library currently contains examples of steels, soft and hard, stainless steels, case hardened and nitrided steels, nickel base alloys, and alloys of aluminium and titanium.

The accuracy of the model depends to a great extent on the amount of detail that can be input in terms of mechanical properties. However, the residual stress can be modelled simply by entering the basic monotonic mechanical properties, viz.:

- Ultimate Tensile Stress
- Elastic Limit
- Young's Modulus of Elasticity
- Elongation at Fracture

Further accuracy of prediction follows if cyclic values can be entered. Based on the input information, the software calculates the cyclic stress-strain behaviour law. If the basic stress-strain cyclic behaviour law is known (e.g. from testing) then the basic values which

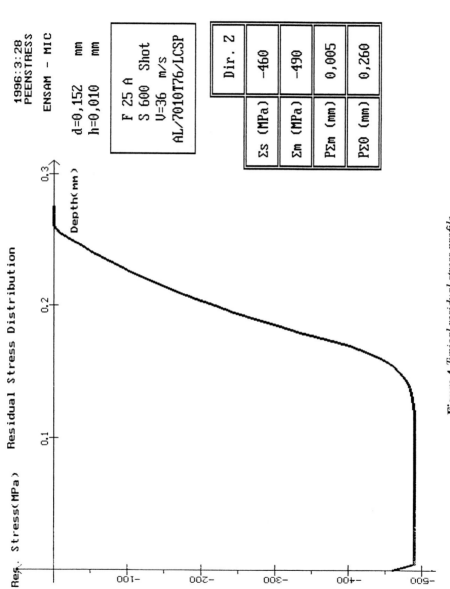

Figure 4 *Typical residual stress profile*

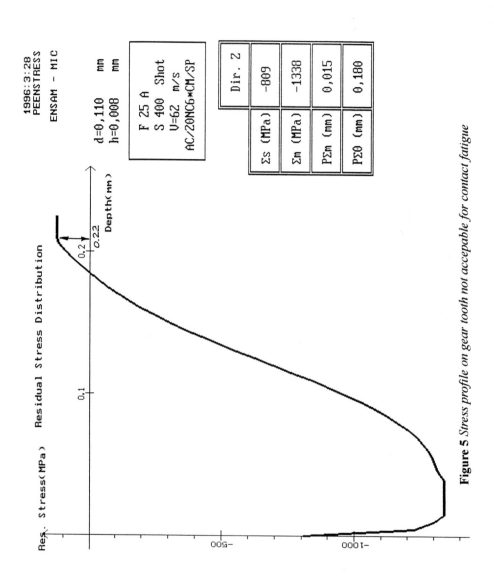

Figure 5 *Stress profile on gear tooth not accepable for contact fatigue*

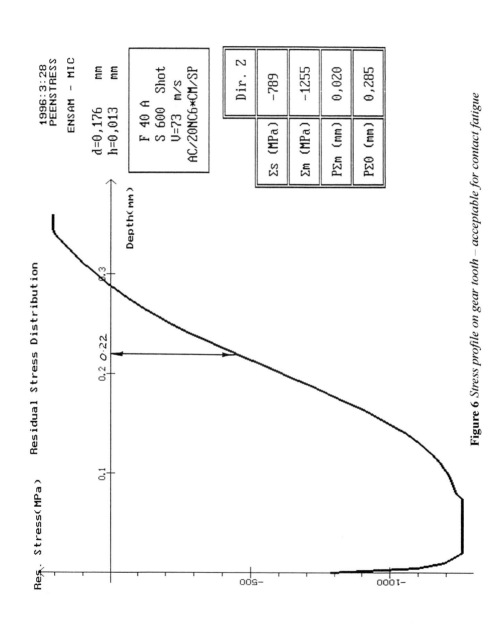

Figure 6 *Stress profile on gear tooth – acceptable for contact fatigue*

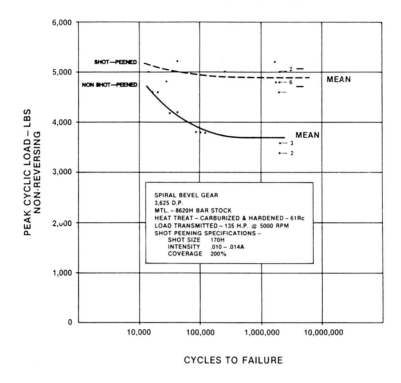

Figure 7 *Increase in fatigue strength of spiral bevel gear*

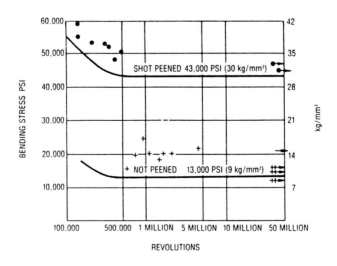

Figure 8 *Safe stress ranges for coil springs (0.207" diameter steel)*

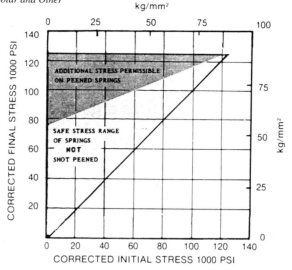

Figure 9 *Fatigue tests on rear axle shafts*

define the behaviour law curve can be added, hence giving the most reliable prediction.

The defining properties of the behaviour law are:

- PO – Initial Hardening Modulus
- SO – Initial Micro-Plasticity
- P – Asymptote Hardening Modulus
    – Asymptote Stress

Both monotonic and cyclic stress-strain curves can be viewed and (Figure 3) shows the monotonic curve for aluminium alloy 7010 T76, with a second (higher) curve showing the results of experimental data.

Any shot peening parameters can be modelled against the chosen base material, and there is the opportunity to input the type of Almen strip (A,N or C), the type of shot, e.g. steel, glass etc., the Almen intensity and the shot size.

Similarly, a number of component shapes can be modelled, since shape, and particularly thickness will affect the stress distribution. The choices are of a flat plate of infinite thickness, a plate of selected thickness, tubular sections of chosen thickness and holes.

A typical printout of a residual stress profile is shown in (Figure 4). The shot peening parameters are reproduced, and as well as a graphical representation of the residual stress profile, the principal points are shown in tabular form, these are the residual stress at the surface, the depth beneath the surface of the peak compressive stress and the depth of compressive stress, and the magnitude of the peak compressive stress.

The theoretical approach was checked, and indeed developed in part with the assistance of many shot peened samples in a large variety of materials. Measurements of residual stress were made using XRD, neutron diffraction and hole-drilling techniques, and good agreement was shown to exist between modelled predictions of peak stress and measured values.

The value of a residual stress modelling tool is, primarily in the understanding of the interaction between the shot peening parameters and the peak value and depth of the resultant compressive residual stress. This is best exemplified by reference to Figures 5 and 6 in which

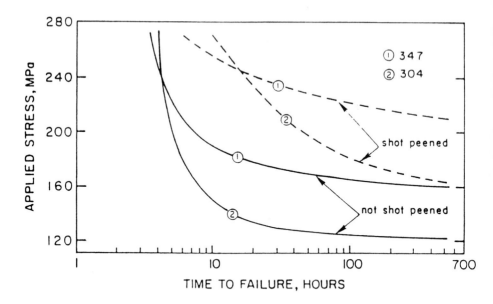

Figure 10 *Effect of shot peening with glass shot on the times to failure of types 304 and 347 stainless steels in boiling 42% magnesium chloride solution*

a fairly commonly used set of shot peening parameters was chosen to enhance the resistance to tooth root bending fatigue of a gear. Figure 5 shows that the neutral stress position occurs at a depth of 190 micron. However, the calculations of the maximum Hertzian contact stress showed a peak of -500MPa at a depth of 220 micron, i.e. in the tensile area. This enabled a higher shot peening intensity to be applied, with the result that the depth of compressive stress was extended to 305 micron (Figure 6).

4 SHOT PEENING APPLICATIONS

As mentioned in the introduction, shot peening has a number of effects on the surface to be treated, and each gives rise to a group of possible beneficial applications for the process.

4.1 Mechanical Fatigue

The main applications undoubtedly arise from the provision of a uniform "layer" of compressive stress in the surface of the component. A number of benefits arises from this, but most commonly it is the improvement in fatigue limit of components which are subjected to cyclic tensile stresses. The improvement arises from the simple premise that any applied tensile stress will be reduced, at the surface, by the magnitude of the residual compressive stress induced by shot peening, or to put it another way, the mean stress will be reduced by the value of the residual stress. Mechanical fatigue, therefore is the main problem for which shot peening is chosen to combat, and is equally applicable to springs, gear teeth, shafts and axles and in fact any engineering component which is subject to cyclic mechanical stress. Typical improvements to be expected from shot peening gears against root bending fatigue range from 25 to 50%[4],(Figure 7), the safe working stress of compression springs can be extended by up to 50%[5,6](Figure 8), and the fatigue limit on automotive rear axle shafts has been increased by over 300%[7],(Figure 9).

4.2 Stress Corrosion Cracking

Another phenomenon which has surface tensile stresses as one of its contributory elements is Stress Corrosion Cracking (SCC) or Environmental Cracking. In a large number of materials, and unfortunately many of which are chosen for their inherent corrosion resistance, the combined effect of a corrosive species and a sustained tensile stress will result in the very rapid deep fissure corrosion attack which is characteristic of SCC. The threshold level of applied stress to drive this mechanism can be as low as 25% of UTS, and seriously affects, for example, austenitic stainless steels in chloride solutions. Shot peening, by "reversing" the surface stress can prevent or delay the onset of this dangerous process. It is most regularly applied to vessels in the chemical industry, and it is well worth noting that in many cases can be used instead of the very costly business of thermal stress-relieving on very large vessels. Furthermore, no matter how well thermal stress relieving is carried out, no account can be taken of applied stresses arising from thermal, pressure and dead weight loading[8] (Figure 10).

4.3 Fretting, Wear and Galling

These phenomena are preventable by the work hardening effect, and refinement in surface finish accomplished by shot peening and less, or nothing, to do with residual stress.

Fretting occurs when relative motion of low amplitude and high frequency exists between two components in close contact. The phenomenon is characterised by the formation of fine particles of oxides on the mating surfaces. These particles are extremely hard, and therefore abrasive and lead to scoring, pitting and even surface fatigue, as well as contaminating lubricants. Fretting fatigue can be completely cured by shot peening[9] and its effect is not dissipated by the application of heat, as is residual stress.

Wear resistance can be improved by the "pre-hardening" of components by the plastic cold work effect of shot peening. This is effective on such materials as Stainless and manganese steels, nickel alloys and stellite. Surface hardness on stellite, for example, can be increased by 50% by repeated cold work applied by shot peening[10].

Galling[11], like SCC, can be a severe problem, and is often overlooked by designers. It is caused by very strong adhesive forces between mating surfaces, generally of like materials, and in particular 300 and 400 series stainless steels, 17-7PH steels, Titanium 6/4 and most aluminium alloys. Galling occurs most frequently on threads and on running fits between like materials in such equipment as valves, where the problem can be between the stem and bushes or between the gate and its seats. Shot peening is carried out on one of the mating surfaces – generally the easiest – such as the male thread. Although quantification of the improvement (and indeed the phenomenon) is difficult, the application remains one of the most effective in the entire scope of shot peening. In order to ensure the best results it is a good general rule that after shot peening the micro hardness of the mating surfaces should differ by around 100 HBN.

4.4 Shot Peen-Forming

There are applications too numerous to mention in this shot paper, but shot peen forming, although applied to a relatively small family of parts, has become one of the most important applications for the process.

In the same way that the Almen calibration strips curve when peened only on one side, the same applies to any other material – this enables controlled curvature to be induced in complex and very large parts without the need for moulds and formers, without the application of heat and external bending loads and at the same time provides a uniform residual stress against subsequent cyclic tensile stresses. For these reasons shot peening has become the standard means by which the airfoil curvature of aircraft wings, and other aerodynamic surfaces, is achieved[12].

5 CONCLUSIONS

From the beginning, shot peening has been used as a remedial technique, often with only experience and testing to verify its effectiveness. Applications which benefit daily from the process encompass a wide range of industry, from automotive and aerospace through to petrochemical, oil exploration and production, mining and agriculture. Our understanding of the process, and more recently the means with which to model its effects, is encouraging the trend seen in designers to use the benefits of shot peening as a design tool to enable material and weight savings, to accomplish greater life and power throughput of an existing design and to avoid some of the pitfalls imposed by material selection or availability detailed above.

References

1. Shot Peening Applications Manual. Metal Improvement Company Inc. 1989, p. 60.
2. H. Guechichi, 'Modelling of the residual stresses induced by shot peening', ENSAM 1986.
3. M. T. Khabou, L. Castex, G. Ingiebert, "The effect of material behaviour law on the theoretical shot peening results". Vol A/Solid, 9, No 6, 1989, p. 537.
4. J. Daly, 'Boosting Gear Life Through Shot Peening', *Machine Design*, May 12 1977.
5. F. P. Zimmerli, Metal Progress, 1952, **67**, 97.
6. 'The Mainspring', Published by Associated Spring Corporation, Bristoi, Connecticut, February 1951.
7. O. J. Horger, and C. Lipson, 'Automotive rear axles and means of improving their fatigue resistance', American Society for Testing and Materials. Technical Publication No 72, 1947.
8. 'The Application of Controlled Shot Peening for the Prevention of Stress Corrosion Cracking', A Technical Review, Metal Improvement Company, 1991.
9. R. B. Waterhause, and D. A. Saunders, *Wear*, 1979, **53**, 381.
10. General Electric Company, Electric Motor Division.
11. G. A. Sprague, 'The New Packard Straight Eight Engine', Presented to Society of Automotive Engineers, June 1948.
12. C. F. Barret, "Peen Forming", 'Tool and Manufacturing Engineers Handbook', 1984, Special Forming Methods.

2.4.2

Applications of High Solar Energy Density Beams in Surface Engineering

A. J. Vázquez[1], G.P. Rodríguez[2] and J. J. de Damborenea[1]

[1]CENTRO NACIONAL DE INVESTIGACIONES METALÚRGICAS, CENIM-CSIC, AV. GREGORIO DEL AMO, 8, 28040–MADRID (E), SPAIN

[2]ETSII, CAMPUS CASTILLA-LA MANCHA UNIVERSITY, CIUDAD REAL, (E), SPAIN

1 INTRODUCTION

Solar Energy (SE) is the oldest source of energy use by the Mankind. It as always available and can thus be used immediately. Probably one of the oldest applications was for drying fruits or vegetables which is still currently applied. Water desalination has surely been the most extensive industrial application of SE and it is still being used at the moment. House building took in consideration of using direct Solar heating in facets of the houses facing south or to protect them against excessive heating by painting the walls white as can now be seen in Mediterranean countries. Building houses with adobe is currently being done. Air conditioning was extensively used favouring water evaporation in porous bricks or in Mountains and recently applied in Seville in a more sophisticated way, but on the basis of the same principle, in the World Fair in 1992. All these examples serve only as an indication of how SE applications are linked to human activity since the oldest times.

When fire was discovered SE applications were shifted to a second position because of the greater power of fire. Since the industrial revolution other sources of energy easier to control and more powerful were applied, and SE applications remained only as a real source of energy in very simple applications in agriculture and in sea water desalination to obtain salt.

However, recently the applications of SE have been considered from an industrial point of view. The increasing worry about the extensive use of non-renewable fuel, world pollution and, more recently, the problem of CO_2 emissions in industrial countries has lead to the consideration of the possibility of using natural energy resources more extensively in a competitive way as a real alternative.

During the fifties some countries, e.g., Israel, made an extensive use of Solar heating in domestic applications. In Odeillo, near the Pyrenees, the French built a large installation for the concentration of solar energy. The first one was built many years ago near Mont Louis, France, applied to study materials at high temperature. When the more industrialised countries encountered the first oil shortage during the seventies some of them thought renewable energy could be an alternative energy source to apply in an industrial way thanks to recent developments in control technology and new materials.

At the moment a team of a number of developed countries decided to build a testing plant in Almeria, SE of Spain, the Almeria's Solar Platform (Plataforma Solar de Almeria, PSA), to test the possibilities of production of electricity as a real alternative source of energy. Actually the installation works with a wider scope not only on this line but also in others such as solar detoxification, water desalination to obtain water, fine chemicals, materials, etc.

Our first contact with PSA was in 1989 and as a result of a technical visit and the information gathered on the possibilities of PSA in Solar concentration we began to work in this direction and our first work on Surface Modification (SUMO) was published a year later.' Then we promoted the organisation of the first team with Prof West from Imperial College, London, Prof Torrance from Trinity College, Dublin, Prof Roos, from Leuven University, and Dr Damborenea in CENIM, Madrid, and we obtained some funds from EU.

Now, all these groups are active and some others are working under the auspices of Access to Large Scale European Installations and we hope that there will be a starting group devoted to future development of Solar applications.

2 CONCENTRATING SOLAR BEAM FACILITIES

Figure 1 shows the world's solar irradiation distribution. Only southern countries in Europe: France, Greece, Italy and Spain, but also Ukraine are in the group of highest irradiation countries. Nevertheless Solar irradiation is not only a problem of latitude but also a problem of altitude and cleanliness of atmosphere. So there are some solar installations in Köln, Germany, and in Davos, Switzerland, were it is possible to obtain enough high energy density values.

Solar irradiation in most Europe is between 0.5 and 1kW/m^2 and it is possible to concentrate it in order to obtain a power density between 200 and 500 W/cm^2. This power density is similar to that of some lasers and it is high enough to produce surface modification materials.

There are many different ways to concentrate energy. The simplest and cheapest is the utilization of a Fresnel lens (Figure 2). These are commercial polymeric lenses approx, 1–1.5 m diameter. With them it is possible to obtain around 100 W/cm^2 on a spot 8 mm in diameter. Even with this very low power density value it is possible to melt a steel sheet 2 mm thick. In other words one can produce SUMO of small pieces of steel or even one can obtain cladding.

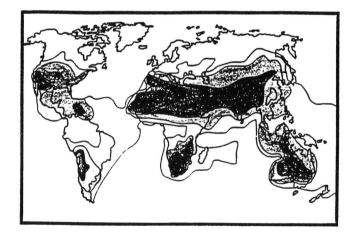

Figure 1 *World solar radiation intensity distribution*

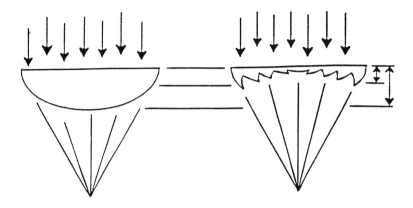

Figure 2 *Scheme of conventional and Fresnel lenses*

A second very simple way of concentrating SE involves using a direct parabolic mirror (Figure 3), normally a multifaceted mirror, facing the Sun and following its track. This mirror concentrates the solar energy in a small spot with a power
density as high as 3 kW/cm^2 on a 5–6 mm diameter spot[2]. Here the sample is face down.

If one or two flat mirrors are in the optic way producing one or two reflections, different designs are possible as can be seen in Figure 4. Here the flat mirror follows the sun track and the surface of the target can in an horizontal position be facing up.

Another larger installation consists of a central tower and a field of heliostats (Figure 5). This is the typical electricity production installation design where the position of the mirrors is parabolic with reference to the central tower. Each mirror, computer controlled, sends the

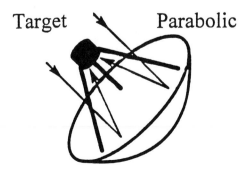

Target   Parabolic

Figure 3 *Single parabolic concentrator*

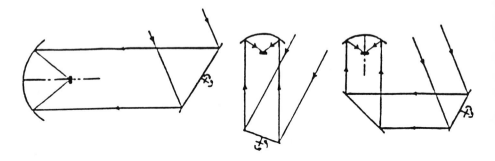

Figure 4 *Scheme of one or two heliostats + Parabolic concentrator*

reflected beam to the same target where there is a superposition of all beams. A power density of about 200 W/cm² is produced in larger surfaces, about 30 cm in diameter. The big problem is that the energy distribution is not constant on all the surface because the spot shape of each reflected beam coming from each mirror depends on its position. However a more uniform energy distribution may be obtained on a 70 mm diameter spot.

Another typical installation, called Solar Furnace, consists of a group of heliostats that direct the solar beam to a multifaceted parabolic mirror. The number and size of heliostats is different according each installation existing in Denver, CO, USA, Almeria, Spain, Odeillo, France, Tashkent, Uzbekistan, etc. The power density may be higher, about 300 W/cm² and the size of the spot is also higher, about 70 mm in diameter in PSA. The main problem of

Target

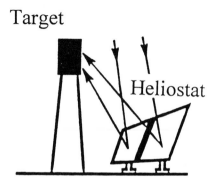

Heliostat

Figure 5 *Scheme of central tower and field of heliostats*

installations with horizontal optical axis is that since the target plane is vertical only non melting treatments are possible, i.e., structural transformation, solid-gas reactions as the treatment of titanium or titanium alloys in a nitrogen atmosphere to obtain titanium nitride or to treat titanium nitride coatings in open air and transform partially the coating to titanium dioxide[3].

To avoid this limitation when the parabolic mirror has a horizontal axis it is possible to add another mirror tilted at an angle of 45 degrees before or after the parabolic mirror so that the concentrated beam can be directed down. With this installation, or with installations where optical axis is vertical samples are in a horizontal plane face up and it is possible to get melting and cladding.

Much higher power densities than those obtained here with these different installations may be obtained using a secondary concentrator[4,5].

Lab experiments may de done with a small Xenon electric discharge lamp because the Xenon lamp has a spectrum similar to that of Solar Radiation. In our laboratory we obtain with a 7 kW power lamp a net power density of 250 W/cm^2. With this power density it is possible to make structural transformation or even melting of steel, nitriding of Ti alloys[7], etc. With a mirror tilted 45 degrees it is also possible to melt the surface of materials without any problem.

3 EXPERIMENTAL RESULTS

To give an idea about the actual possibilities of Solar Engineering Applications on Materials (SEAM) we will present some data from our work and others in literature, where an extensive review is made by Rodriguez-Donoso[8]. Table 1 presents a summary of some installations existing in the world where quite different values of total power and power density are achieved. Table 2 presents a summary of other processes that use high power density beam installations showing maximum power density and total power.

Table 1 *Main characteristics of solar plants*

Site	Type	Power (kW)	Power density W/cm^2
	CRS Central Tower	3362	250
Almería (Spain)	CESA Central Tower	7000	250
	Solar Furnace	60	500
Odeillo (France)	Heliostat + horizontal axis	1000	1600
	Heliostat + verticle axis	6.5	1500
Denver, CO	Heliostat + horizontal axis	1000	250
Albuquerque, NM	Central Tower	3350	260
(USA)	Heliostat + parabolic	1600	300
Iraq	Fresnel lens	0.08	
	Central Tower	1000	1000
Tashkent (Uzbekistan)	Heliostat + vertical axis	3350	800
	Heliostat + horizontal axis	1600	100
Israel	Central Tower	2900	
	Heliostat + parabolic	16	1100
Gungzhou (China)	Direct parabolic	1.7	3000

Table 2 *Power and power densities in different beam processes*

Process	Power (kW)	Power density (kW/cm²)
Concentrated solar beam	500	3 10*
Arc lamp	300	355
Plasma	500	100–1000
Electron beam	300	1000
Laser	100	1–10

*With secondary concentrator

3.1 Structural Transformation

In 1982 in China tests were done on hardening of steels 0.8 % in carbon, with a parabolic concentrator 1.56m in diameter. The power density used was up 3 kW/cm² and the focal spot size was 6 mm in diameter. They used steel cylindrical samples 30–40 mm in diameter and 1.3–5 mm of thickness with a high carbon content. With a solar radiant intensity of 850 W/m² the treatment time was 1s, and with 445 W/m² the treatment time was 7s. Hardening was 900 HV 0.5 on various spots up to 5mm in diameter and 0.5mm depth (Figure 6).

Treatment of larger surfaces was performed with several tracks. The overlapped zones underwent tempering as can be seen in Figure 7. These data correspond to a treatment of 2 s with a solar radiant intensity of 847 W/m² with an overlapping of 1mm.

Yu et al.[9] present some practical results on the application of a 600 SE treatment to a milling screw. Zhong et al.[10] present the results of the treatment of a piece of a gun machine.

Mayboroda et al.[11] used a faceted parabolic concentrator with a 200–400 W/Cm² power radius (mm) density and compared their results with those obtained in a conventional treatment.

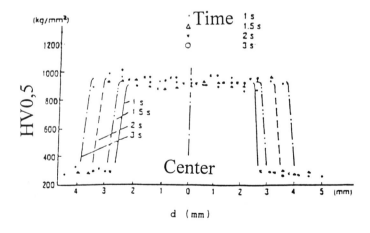

Figure 6 *Hardness through the treated spot*

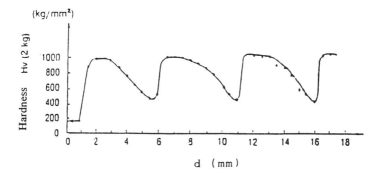

Figure 7 *Tempering effect of several tracks*

Pitts et al.[12] work in Sandia with an heliostat and a multifaceted parabolic concentrator. Maximum power density was 300 W/cm². Samples 50 mm in diameter and 1.6 mm thickness of tool steel were heated to 1523 K. Hardening of the sample is presented in Figure 8. Stanley et al.[13] did several tracks on a sample 610 x 150 mm and 13 mm thickness with 200 W/Cm² power density. They obtained a 20 mm wide track with a treatment depth of 1–2 mm with a maximum hardness of 600 HV.

Vàzquez et al.[1] performed successfully preliminary work involving the treatment of different steels. In a subsequent programme[14] samples 35mm in diameter and different heights were treated in a central tower and heliostat field installation and a hardeness of 600 HV(C < 0.38 %) and 900 HV (C = 1.0 %) were obtained. Rodriguez et al. obtained with a F. 1202[15] steel similar hardness than with a conventional treatment. More recently[16] AISI 4140 steel was treated in a central tower and heliostats field installation and 700 HV was obtained with a very sharp hardness profile (Figure 9). With a Fresnel lens Rodriguez et al.[17] with only 80 W/cm² reported that hardening of steel is possible, but also easy changes in temperature through the modification of insulation. In some cases, samples were treated with a black paint to enhance the absorption[18].

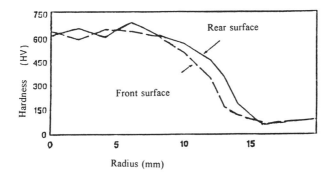

Figure 8 *Hardness in steel sheet*

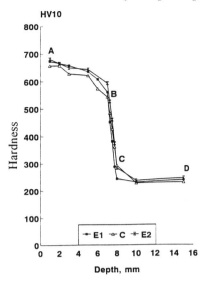

Figure 9 *AISI 4140 hardness profile*

3.2 Surface Coating or Modification

Other applications of SEAM are chemical modification and cladding. Korol et al.[19] obtained coatings 50–150 microns with a parabolic mirror 2m in diameter with a composition TiC-Ni-B and WC-NI-B on a base of stainless steel and Ti alloy. The treatment was made in 1-3 s with power densities of 1200–600 W/Cm².

Yu et al.[20] laid a cast iron sample with WC and melted it with SE. They got an alloy coating 0.5–1 mm thickness without any inclusion or porosity and with very high hardness values. A large coated surface was obtained with several tracks with overlapping without any diminution in hardness because of the high tempering resistance of tungsten (Figure 10).

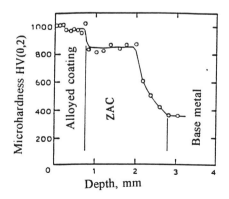

Figure 10 *Hardness of WC coating on cast iron*

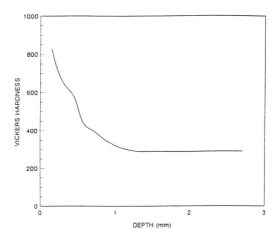

Figure 11 *Hardness of a Ni-B cladded coating*

Ivaschenko et al.[21] treated sintered tungsten carbide and cobalt (VK8) and powder tungsten carbides and titanium carbides sintered with cobalt (T15K6) in an SE installation with 100–200 W/cm^2. The life of the tool was 1.7 to 4.5 longer. With VK8 titanium oxicarbide was formed and with T 1 5 K6 titanium carbonitride was formed.

Stanley et al.[13] melted Ni and Cr composite powders and Ni and Al composite powders on tool steels with 100–200 W/cm^2 during 4–15 s. In other paper they show very good adhesion of the melted powder to the base material. Also they did CVD with SE and produce coatings of Ni$_3$Al, 1 micron in thickness starting with mixtures of Al and Ni powders to produce SHS. Lewandowsky[23] and Rawers[24] synthesize films and coatings.

Franck et al.[3] changed a previous coating of titanium nitride in rutile (TiO$_2$) in the central tower and heliostats field installation of PSA.

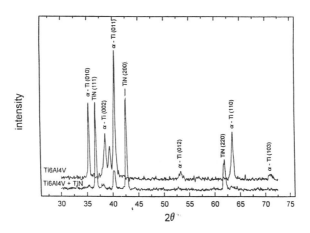

Figure 12 *Spectra of TiN coating on a Ti6Al4V base material*

Fernández et al.[25] obtained a cladding of Ni-B on a plain carbon steel in the PSA Solar Furnace with very good hardness values (Figure 11). Rodriguez et al.[7] also obtain a high quality coating of TiN with a Xenon lamp, as can be seen in the comparison of the both spectra of base material and coating (Figure 12) as a first step to produce these coatings in the comparison of the both sectra of base materials and coating (Figure 12) as a first step to produce these coatings in the PSA facilities.

3.3 Welding

Kaddou et al.[26] welded 3 mm thickness pieces of bronze and copper with a Fresnel lens 360 mm in diameter with a 80 W/Cm^2 power density.

3.4 Modelization

Yu et al.[9] presented a calculation of the surface temperature and subsurface zone in samples with the hypothesis of unidirectional heating flow.

Yang et al. [27] also develop a computer program to calculate the temperature of samples treated with SE.

Ruiz et al.[28] present a computer program to be applied in surface modification. On the basis of SE parameters and base metal characteristics net surface heating may be calculated and temperature and hardness profiles of quenched sample estimated.

4 CONCLUSION

As it can be seen SUMO with Solar Energy is a real possibility and not a chimera nor a fairy tale. Certainly a lot of work will be done before these applications are included in industrial processes, but the time required to achieve this success depends on the effort used in the study of this application.

Our next effort will be in the field of cladding with an installation where the power density would be high enough to modify the surface maintaining the structure of the base material. Another line of study will be the development of a fluidized bed to make the heat treatment of steels and other metallic pieces. We hope that in this year this equipment will finished and tested. The control of the temperature of the fluidized bed will be compensated with simultaneous electric heating to avoid fluctuations when the flux input will decrease as a consequence of changing of atmospheric conditions. We hope this parallel electric heat input will be lower than 10 % of total input, so a saving in energy of 90 % can be obtained.

The possibilities of accurately achieving heat input control with electronic devices will be another opportunity to implement this kind of equipment to use solar energy. The quality of mirrors, the life and the ever decreasing costs[29] lead us to think that Solar Energy Application to Materials SEAM, will be a real alternative, first in Research applications and later in Industry.

References

1. A. J. Vázquez, J. J. Damborenea, Rev. *Metal. Madrid*, 1990, **26**, 157.
2. Z. K. Yu, Q. Y. Zong, Z. T. Tam, *J. Heat Treating*, 1982, **2**, 344.

3. M. B. Franck, M. B. Blanpain, B.C. Oberländer, J. P. Celis, J. R. Roos, *Solar Energy Materials and Solar Cells*, 1993, **31**, 401.

4. J. J. O' Gallagher, R. Winston and A. Lewandowsky, AES Solar 93 Conf, Washington. D. C. April, 1993.

5. A. Lewandowsky, C. Bingham, J. O'Callagher, R. Winston, D. Sagie, *Solar Energy Materials*, 1991, **24**, 550.

6. G. P. Rodriguez, A. J. Vázquez and J. J. de Damborenea, *Mat. & Des.*, 1993, **3**, 169.

7. C. Rodriguez, I. Garcia, J. J. de Damborenea, A. J. Vdzquez, CAM Progr. Reports 1996, Madrid.

8. G. P. Rodriguez-Donoso, PhD Thesis, Univ. Complutense Madrid, 1994.

9. Z. K. Yu, Q. Y. Zong and Z. T. Tam, *J. Heat Treating*, 1982, **3**, 120.

10. Q. Zhong, Z. Tam and M. Cao, *J. Heat. Treat. Met*, 1986, **6**, 15.

11. P. V. Mayboroda, V. V. Pasichniy, N. G. Palaguta, A. I. Stegniy, V. G. Krivenko. *Metalloved Term.,Obrab. Met.*, 1986, **1**, 59.

12. J. R. Pitts, J. T. Stanley and C. L. Fields, Proc. 4th Int Symp. on Res. Dev. and Appl. of Solar Thermal Technol., Sta. Fe, NM., USA, (6),1988.

13. J. T. Stanley, J. R. Pitts, C. L. Fields, 'Protective Coatings: Processing and Characteristics', R.M. Yacizi, 1989, p. 43.

14 A. J. Vázquez, G.P. Rodriguez and J. J. de Dainborenea, *Solar Energy Materials*, 1991, **24**, 751.

15. G. P. Rodríguez, V. López. A. J. VAzquez, J. J. Damborenea, 'Solar Engineering', A. Kirkpatrick, Ed. ASME, N.Y., 1993, p. 325.

16. G. P. Rodríguez, V. López, A. J. Vázquez and J. J. Damborenea, *Solar Energy Materials and Solar Cells*, 1995, **37**, 1.

17. G. P. Rodríguez, A. J. Vázquez, and J. J. Damborenea, *Mat. Sci. Forum*, 1994, **163–164**, 133.

18. A. Torrance, First Progress Report on Contract 67/90: Large Installation Program, Dublin Trinity College, 1991.

19. A. A. Korol, E. A. Korol, E.A. Kasich-Pilipenko, I. E. Verkhovodov, P.A. Dvevnyakov and V. Kadryov., *Poroshk. Metall.*, 1983, **4**, 39.

20. Z. K. Yu and J.T. Lu, *Surf Eng.*, 1987, **3**, 41.

21. L. A. Ivaschenko, G.V. Rusakov, V. V. Pasichniy, A. I. Stegnii, S. S. Ponomarev, *Sov. Powder Metall. Ceram.*, 1989, **28**, 64.

22. J. R. Pitts, C. L. Fields, J. T. Stanley, B. L. Pelton, Proc. 25th Intersociety Energy Conversion, Eng. Conf, AlChE, N.Y., 6, 1990,-p. 262.

23. A. Lewandowsky, *Mat. Tech.*, 1993, **8**, 237.

24. J. C. Rawers, D. E. Alman, A. Lewandowsaki, A. V. Petty and J. Pitts, *J. Mat. Sci. Letters*, 1994, **13**, 1608.

25. J. Fernández, B.J., Martínez, V. López A.J. Vázquez, *ISATA Conf*, Florence, 1996 (in press).

26. A. F. Kaddou, A. Abdul-Latif, *Solar Energy*, 1969, **12**, 377.

27. Y. Yang, A. A. Torrance, J. Rodríguez, *Solar hardening.. experimentalprediction:* Final Report on the Program Access to Large-Scale Scientific Installations, Almeria, PSA, July, 1993.

28. G. P. Rodríguez-Donoso, J.R. Ruíz, B.J. Fernández, A.J. Vázquez-Vaamonde, *Mat. and Des.*, 1995, **16**, 163.

29. M. Sanchez González, PhD, Thesis, Universidad Complutense Madrid, Oct. 1995.

2.4.3

Processing and Properties of Thermoplastic Coatings on Metals

D. T. Gawne and Y. Bao

SCHOOL OF ENGINEERING SYSTEMS AND DESIGN, SOUTH BANK UNIVERSITY, LONDON, UK

1 INTRODUCTION

Despite the many advantages of engineering polymers, they are inferior to metals in some applications. In particular, polymers have a low strength relative to metals and also a much greater thermal expansion coefficient. These two physical properties result in a low load-bearing capacity and poor dimensional stability which limits the use of bulk polymers, especially where high tolerances under load are required. However, polymers as coatings on metals have unique advantages. The load-bearing ability of the component is provided by the underlying metal and the dimensional changes of the polymer are small owing to the small cross-section thickness of the coating . Polymer coating on metal components, therefore, have the potential of upgrading quality by providing corrosion resistance, reducing the demand for lubrication, absorbing vibrations and reducing noise levels, reducing friction and some limited ability to embed abrasive particle or debris. The traditional methods of applying polymers on engineering components and structures are by painting and powder coating. Painting involves dissolving the polymer in a solvent, in order to provide a binder for the pigment and facilitate coverage of the substrate by application in the liquid state. Painting is a low-cost and highly developed technology but is facing increasing problems, including emissions caused by the solvents in wet paints and the rising cost of waste disposal. Painting also has difficulties in the requirement for drying, which leads to substantial downtimes, as well as the control of coating thickness and the limited quality of the coatings.

Powder coating involves the formation of a polymer coating on a component by the application of a dry powder to its surface in conjunction with stoving at an elevated-temperature. Polymer powder coating has been developed substantially over the last fifteen years: it now occupies 5–10% of the total industrial painting and is a possible alternative to painting in many cases[1,2]. Polymer powder coating processes have several advantages over conventional painting. One of the most significant is that they are more environmentally acceptable. There is no solvent to cause air pollution and less make-up air is required in the stoving ovens than for solvent-based paints, where the volume must be high enough to keep the volatiles below explosive concentrations. In addition, powder coating involves relatively little waste as stray powder or overspray can be collected and re-used[3]. For example, in the application booth for electrostatic spraying, the overspray powder can be removed from the air, the powder recycled and the air returned to the building. In contrast, the disposal of sludge produced by overspray in conventional painting operations is becoming increasingly expensive.

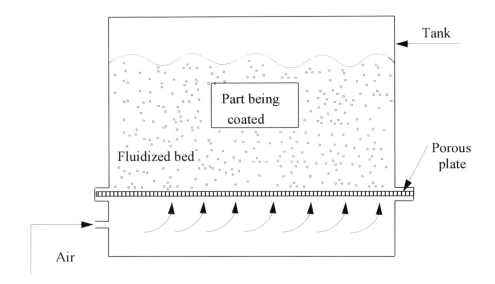

Figure 1 *Schematic of fluidized bed*

2 CONVENTIONAL POLYMER POWDER COATING PROCESSES

Several techniques employing powdered polymers are available for applying polymer coatings to various components. The principal ones in commercial use are electrostatic spraying and fluidized bed dipping.

2.1 Fluidized Bed Dipping

The fluidized bed dipping is the oldest powder coating process and originated in Germany in late 1950s[4]. In this process, the coating is carried out using a fluidized bed system. The schematic diagram of a fluidized bed is shown in Figure 1. The fluidized bed consists of two compartments separated by a porous plate. Cold compressed air is forced through the porous plate to the compartment containing powder, causing fluidization in which the powder is suspended in a fluidized state and behaves like a liquid. The substrate to be coated is preheated above the melting point of the polymer powder in an oven and then dipped into the fluidized bed of the powder. The powder melts and forms a coating. The fluidized bed dipping process can produce polymer coatings with the thicknesses of 200 to 2000 μm, but it is difficult to apply thinner coatings of less than 200 μm. In addition, this process is seldom used for large objects because of the size limit of the fluidized bed and powder reservoir required.

2.2 Electrostatic Spraying

Electrostatic spraying gained commercial acceptance in early 1960s. In the electrostatic spray process, the coating powder is withdrawn from a fluidized feed hopper by clean, dry

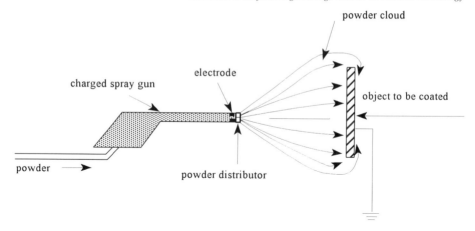

Figure 2 *Schematic diagram of an electrostatic spray gun*

compressed air and electrostatically charged in the high voltage corona field of a spray gun[3]. The charged particles are attracted to the earthed metal substrate which is to be coated and adhere to its surface by electrostatic attraction. A typical electrostatic spray gun is shown in Figure 2. The coated substrate is then placed in an oven and the particles are melted and fuse together to form a continuous coating. The electrostatic spray process can be used to coat a wide variety of objects with various configurations. When powder is electrostatically applied to parts at ambient temperature, the coating thickness tends to be self-limiting (20–75 μm)[3] since the substrate becomes insulated by the powder after the powder is deposited on the surface of the substrate. Modest increases in coating thicknesses can be obtained by using a preheated substrate, which relies on the residual heat to attract more powder after insulation has occurred. Airstatic spraying is similar to electrostatic spraying but has the added safety factor that it is spark-free: the powder is charged by friction using on ion generator[7]. Flock spraying consists of blowing the powder with compressed air on to a preheated substrate. The process is less controllable than electrostatic spraying and has a relatively low deposition efficiency[7].

3 THERMAL SPRAYING

Thermal spraying is a process of particulate deposition, in which particulate materials are melted to form droplets and then propelled on to a substrate surface where the individual particles impact, spread and solidify into splats. Repeated scanning of the substrate with the spray gun enables the splats to cover the surface and incrementally to build up a coating. There are three basic thermal spraying coating processes: combustion spraying, plasma spraying and electric arc spraying. Each process has its own advantages and field of application.

Combustion spraying uses the heat of combustion of a fuel gas (acetylene, propane, etc.) and oxygen (or air) to melt the coating material. The two types of combustion spraying are conventional flame spraying or low-velocity oxy-fuel spraying (LVOF) and high-velocity oxy-fuel spraying (HVOF). The process is portable and versatile. Work can be performed on-site

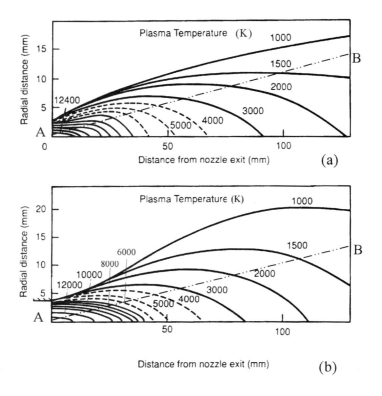

Figure 3 *Temperature distribution of (a) a nitrogen-hydrogen plasma and (b) an argon hydrogen plasma*

and large objects may be coated without the need for large tanks or ovens which are generally necessary for fluidized bed dipping and electrostatic spraying. Small components may also be coated easily. LVOF spraying has been used to produce polymer coatings ranging from 250 μm to over 1 μm in thickness. However, the process depends on greatly on operator skill and surface finish of the coating may be inferior to that produced by other techniques. The polymer coatings may suffer degradation from overheating under oxidizing atmospheres. Although LVOF coatings have limited quality, they have been used in a number of industrial applications for decoration, corrosion and wear resistance, seals and abradable coatings[8-11]. HVOF spraying generates much higher gas velocities (700–2000 ms^{-1}) which should provide higher quality coatings and the process has been applied successfully to spray aluminium-polyester abradable coatings[12].

Electric arc spraying uses a wire or rod as the coating material, which is automatically fed into a flame where it melts and is blasted on to the substrate by compressed air. The heat is supplied by d.c. arc struck between the two metal wires. Combustion spraying processes include combustion wire spraying which uses wire or rod as a feedstock material. No work has been reported on the wire-spraying of polymer coatings.

Plasma spraying also uses high temperature gas jets to heat particulate materials but the jet is formed by heating inert gases in an electric arc not by combustion. The gases are ionized,

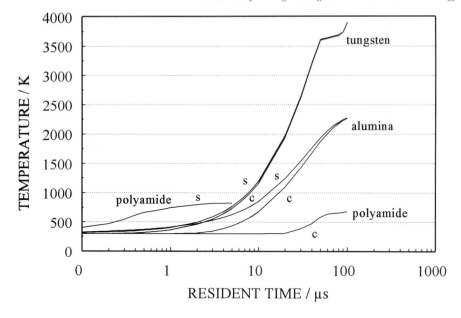

Figure 4 *Calculated temperature rise of tungsten, alumina and polymide particles in a thermal plasma (s and c refer to the surface and centre of each particle)*

and then used as a source of thermal and kinetic energy to melt and propel the powder particles on to a substrate to produce coatings. Unlike combustion flame spraying, the heat source is an inert plasma gas, which suppresses the degradation/oxidation of the particles in flight. There are three main types of plasma spray process: air plasma spraying(APS), carried out in air, vacuum plasma spraying (VPS) or low pressure plasma spraying (LPPS) carried out in a chamber under reduced pressure, and atmosphere-controlled plasma spraying carried out in a chamber containing an inert (IPS) or reactive gas (RPS). Figure 3 gives a typical temperature profile of a APS plasma beam[13]. It shows that the temperature at nozzle exit is over 10 000°C, but falls to below 3000°C at a distance of 100mm from the nozzle. With such a broad temperature profile the plasma spray process offers a wide range of materials which can be manipulated from low melting point polymeric materials to refractory metals and ceramics. Plasma spraying also provides particles within the jet a high in-flight velocity. The beam velocity of a plasma jet can reach to 800ms^{-1} or even more, depending upon the geometry of the torch nozzle[14], which are much greater than those in LVOF spraying (80–100 ms^{-1}). The high beam velocity gives the powder particles a high in-flight particle velocity, which promotes the coherence of individual splats, the coating adhesion of coating to substrate and reduces the porosity.

4 THERMAL SPRAYING OF POLYMER COATINGS

(a) Heating of in-flight polymer particles in thermal plasmas

Polymeric materials possess low thermal conductivities, low melting temperatures and decomposition temperatures compared with metals and ceramics. The thermal character of polymers affects their behaviour on heating in the plasma. The temperature profiles of in-

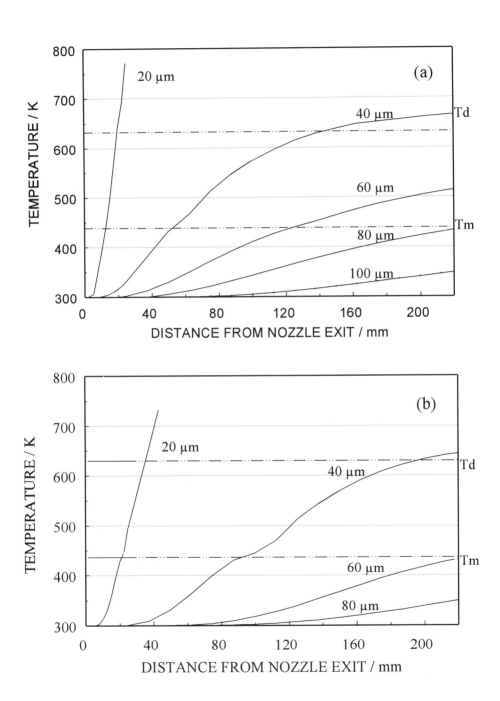

Figure 5 *The effect of particle size on calculated particle temperature: (a) at half-radius and (b) at centre as a function of spraying distance*

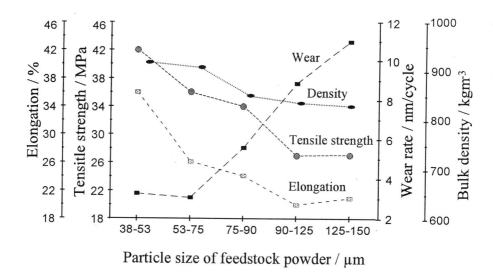

Figure 6 *The effect of feedstock particle size on the quality of the plasma sprayed polymid 11 coatings*

flight particles of a typical metal, ceramic and polymer were calculated[15] and the results are shown in Figure 4, in which the temperature profiles of in-flight particles are expressed as a function of residence time of the particles in the plasma jet. It can be seen from Figure 4 that

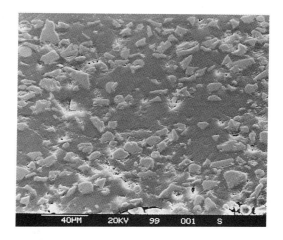

Figure 7 *SEM micrograph of polished cross-section of alumina-polymid composite coating*

Figure 8 *SEM micrograph of polished cross-section of glass-polymid coating with glass particle size of 0–20 μm*

the temperature at the particle centre rises more slowly than that at the surface for all materials. The effect is particularly marked for polyamide, where the temperature at surface has reached its thermal decomposition temperature (upper limit 830 K) while the temperature at the centre is still 300K. A notable characteristic is that the surface temperature of the polyamide particle rises significantly more rapidly than that of the tungsten and alumina particles, while its centre temperature rises significantly more slowly. This results in a large temperature gradient developing within the polyamide particle. In contrast, the temperature difference between surface and centre of the alumina particle is relatively small, while that of tungsten is almost negligible relative to the polyamide particle.

The development of large temperature gradients is related to the internal heat conduction developed within the particle. The conductive heat flux is proportional to the thermal conductivity and temperature difference according to Fourier's law[16]. Polyamide has a low thermal conductivity of 0.29 $Wm^{-1}K^{-1}$ compared with 110 $Wm^{-1}K^{-1}$ for tungsten, and 6.3 $Wm^{-1}K^{-1}$ for alumina. In addition, it has a low thermal decomposition temperature (upper limit 830K) above which its surface temperature cannot rise.

The thermal conductivity of most engineering polymers[17] is less than 0.5$Wm^{-1}K^{-1}$. As a result, high internal heat conduction resistance leading to the development of large temperature gradients within particles should be expected to be a unique characteristic of polymers during thermal spraying. The heating of polymer particles in a thermal plasma is thus limited by internal heat conduction, whereas the heating of high conductivity materials, such as metals, tends to be controlled much more by the heat supplied by the plasma. As a result, the heating of polymers is particularly dependent upon the particle size of the feedstock powder.

(b) Effect of particle size on coating quality

The quality of an engineering coating depends upon its properties in relation to the fitness for the intended application, but generally a low porosity, coherence and adhesion to the

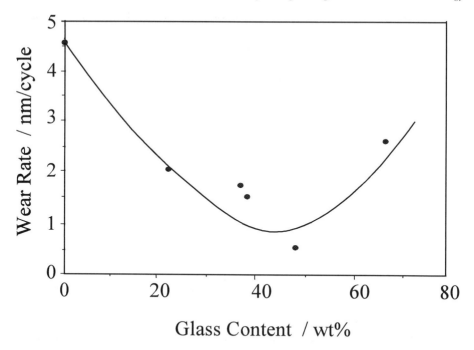

Figure 9 *Effects of glass content on equilibrium wear rate of glass-polymid composite coating*

substrate are desirable characteristics. The quality of thermally sprayed coatings is dependent upon the state of the particles, particularly the degree of melting, immediately before impact with the substrate. The temperature of the particle determines the degree of melting and decomposition of in-flight particles, together with their melt viscosity and flow on impact with the substrate.

Heat transfer analysis is of great value in theoretically predicting the temperature profiles and history of the particles and deposited layers. A process model based on heat transfer theory developed by the authors[15] has been applied to calculate the temperature profiles of polyamide particles of various sizes in plasma spraying. Figure 5 shows the calculated temperature at the half-radius (Figure 5a) and centre (Figure 5b) of polyamide particles with various diameters as a function of spraying distance. The particles were immersed in an argonhydrogen plasma of 22 kW arc power with in-flight velocities of 240ms^{-1}. The calculated results show that the particle size has a crucial effect on the particle temperature: for example, large particles (80–100 μm) at a typical spraying distance of 150 mm were not melted at their centre nor half-radius, while small particles (20 μm) exceed their decomposition temperature (Td) at only a distance of only 30mm, which is, in fact, impracticably short for spraying. The smaller particles have shorter paths for heat conduction from the surface to the centre, and hence require a shorter time and less thermal energy for heating.

The above data predict that the temperature and state of feedstock particles arriving at the substrate depend critically on their size. Experimental work was undertaken to determine the consequences of this finding on the quality of the coatings. A series of polyamide coatings was plasma sprayed using feedstock powders with different particle sizes, and their quality assessed in terms of density, tensile strength, total tensile elongation and wear rate. The tensile measurements were made on free-standing coatings with a 25 mm gauge length and wear was evaluated on a reciprocating ball-on-flat machine with a 12.7 mm stainless steel ball under a load of 20N. The experimental results are given in Figure 6 and indicate that the feedstock particle size has a major effect on quality. The maximum quality was obtained with the 38–53 μm feedstock particle size. This size provides sufficient melting and a low enough viscosity for the polymer to flow readily into the irregularities of the underlying topography, which produces tightly knitted splats and minimizes voidage in the coating.

(c) Polymer matrix composite coatings

Polymers are applied as protective coatings on metals but their usage is restricted by their limited wear resistance. Experience from bulk materials has shown that this limitation can be overcome to a certain extent by the use of fillers. However, the conventional processes of electrostatic spraying and fluidized bed dipping are not amenable to the incorporation of fillers into the coating. Thermal spraying has potential for achieving this objective by injecting two or more types of powders into the flame. The injection points can be selected such that the polymer matrix phase melts in the flame and the filler is introduced further downstream to avoid melting and overheating the polymer.

The authors have deposited a number of polymer-matrix composite coatings on steel by plasma spraying, including alumina-polyamide (PA), glass-PA, magnesium hydroxide-PA and stainless steel-PA[18,19]. Figure 7 shows intimates bonding between the alumina filler and the polyamide matrix in a composite coating, whereas Figure 8 shows close bonding between the matrix and the steel substrate in a glass-polyamide composite coating. The alumina particles have retained their angular shape indicating that they did not melt in the flame. These observations demonstrate that dense deposits with sound bonding can be produced by plasma spraying.

No significant deterioration in the coating-substrate adhesion of the composite coatings relative to pure polyamide was found, which was attributed mainly to the formation of a denuded interface layer next to the substrate[19]. An investigation of the wear of a glass-polyamide coating was carried out and the results are shown in Figure 9. The presence of the glass filler reduces the wear rate of polyamide by almost an order of magnitude, because of its load-supporting action and reduction of adhesive and abrasive wear. However, the results in Figure 9 show that there is a minimum in the wear rate and a corresponding optimum filler content of 50wt%. This is attributed to the fact that fracture initiates by decohesion at the filler-matrix interface and above 50wt% the filler spacing becomes small enough for the stress fields to interact and stimulate cooperative fracture with cracks propagating from one filler particle to the next. The research also investigated the effect of filler size on wear performance with fillers in the range 0–200 μm. The work showed that the wear resistance improved with decreasing filler size with the 0–20 μm size fraction producing the minimum wear rate[19].

5 SUMMARY

1. The potential of polymers as engineering coatings has not been achieved

2. The existing powder coating methods have limitations in the application of polymer coatings.
3. Thermal spray processes have unique advantages for the production of engineering polymer coatings.

References

1. D. T. Gawne and I. R. Christie, *Metals and Materials*, 1992, **8**, 646.
2. A. Matthews, R. J. Artley, P. Holliday and P. Stevenson, 'U.K. engineering coatings industry in 2005', DTI/Hull University, 1992.
3. K. J. Coeling and T. J. Bublink in 'Encyclopedia of polymer science and engineering' Vol.3, J. I. Kroschwitz, ed., John Wiley and Sons 1985, p. 575.
4. J. Gaynor, *SPE J.* 1959,**15**, 1059.
5. W. R. Pascoe, *Mater. Des. Eng.*,1960, **51**, 91.
6. M. J. Day, 'The Welding Institute Research Bulletin'. Dec 1987, p. 409.
7. M. F. Brooks, *Surface Journal International*, 1986, **11**, 26.
8. Metallisation Ltd, *Anticorrosion*, Aug 1988, 19.
9. T. W. Glass and J. A. DePay, 'Protective thermoplastic powder coating specifically designed adhesive polymers'. Proc. 4th Nat. Thermal Spray Conf. Pittsburgh, PA, USA, 4–10 May 1991, p. 345.
10. C. E. Blackinore, *Surface Engineering*, 1987, **3**, 29.
11. P. J. Loustaunau and D. Horton, *Materials Performance*, July l 994, 32.
12. A. R. Nicoll, A. Bachmann, J. R. Moens and G. Loewe, 'The application of high velocity combustion spraying'. Proceed. of the Int. Thermal Spray Conf & Exposition, Orando, Florida, USA, 28 May–5 June 1992, p. 149.
13. M. Vardelle, V. Vardelle, P. Fauchais and M.I. Boulos, *AIChE Journal*, 1983, **29**, 236.
14. L. Pawlowski, 'The science and engineering of thermal spray coatings', John Wiley & Sons, Chichester, UK, 1995, p. 22.
15. Y. Bao, D. T. Gawne and T. Zhang, *Trans. Inst. Metal Finishing*, 1995, **73**, 119.
16. D. R. Croft and D. G. Lilley, 'Heat transfer calculations using finite difference equations'. Applied Science Publ. London, 1977 p. 1.
17. M. F. Ashby, *Acta metall.*, 1989, **37**, 1273.
18. Y. Bao, D.T. Gawne, D. Vesely and M. J. Bevis, *Trans. Inst. Metal Finishing*, **72**, 1994, 110.
19. Y. Bao and D. T. Gawne, *J. Mat. Sci.*, 1994, **29**, 1051.

2.4.4
Ultrasonic Flow Polishing

A. R. Jones and J. B. Hull

DEPARTMENT OF MECHANICAL ENGINEERING, THE NOTTINGHAM TRENT UNIVERSITY, NOTTINGHAM, UK

1 BACKGROUND

A high quality surface finish is required in dies and moulds used in the fabrication of metal, glass and plastic products. In order to ensure not only an optically acceptable surface finish but also to permit ease of release, blow moulding dies require a polished surface with R_a values better than 0.5 μm. Similarly, for reasons of both appearance and integrity, steel dies formed by electro-discharge machining (EDM) require the removal of the thermal re-cast layer.

The majority of the surface finishing processes used on these dies and moulds are currently performed manually. In the case of blow moulding dies, manufactured from aluminium, some surface improvement of easily accessible areas can be carried out with small rotary abrasive wheels. However, the greater part of the work is performed manually using abrasive cloths of successively decreasing abrasive particle size. The process culminates in buffing the mould with impregnated cloth.

Such manual polishing methods have a number of limitations, all of which are a result of operator shortcomings. Firstly, the operator has to be trained in the technique and whilst the approach may be simple for flat surfaces, the polishing of undercuts and radii requires the implementation of special methods, which requires training and experience.

The polishing also requires repeatability. For example, a skilled worker can be required to spend ten days polishing ten complex dies and needs to ensure that, on completion, all the dies are identical. Finally, it is important that, in the achievement of a polished surface, there is no loss of dimensional accuracy. As a result, due to his skills at polishing, a trained and experienced worker becomes an expensive but essential part of the manufacturing process. The process is also extremely time consuming. It has been reported[1] that the polishing time of a mould can represent up to 37% of the total production time of the entire mould. Improvement in the finishing or polishing of the mould or die to lessen the input of a skilled operator or to reduce the processing time has the potential to dramatically reduce the cost of the finished item.

Whilst automated processes, suitable for the polishing of closed dies exist, they are limited in their application.

Abrasive flow machining [2-4] is currently used in industrial applications to successfully and efficiently polish open dies. It uses a mix of abrasive particles suspended in a pliable polymer base which is hydraulically powered over the surface to be polished. The process is however normally restricted to the polishing of workpieces which are tubular or hollow in form with an entry and exit for the flowing mix.

Controlled material removal can be achieved by the use of ultrasonic machining[5-7]. In this process, the workpiece material is removed by the high frequency hammering of abrasive particles, in the form of a waterbased slurry, into the workpiece surface. By the use of suitably formed tools complex shapes can be created in the workpiece. As a potential method of polishing closed dies, this machining method has the disadvantage that it is normally used for fixed location drilling or workpiece indentation.

It was considered that if the processes of abrasive flow machining and ultrasonic machining could be combined in such a fashion that the abrasive flow mix could be energised ultrasonically, then there would exist the potential of a process capable of finishing and polishing dies and moulds. It was with this background that a major Brite Euram Project was set up. Its objective was to develop a method of polishing closed dies.

2 THE ABRASIVE FLOW MACHINING (AFM) PROCESS

Abrasive flow machining is a non-traditional machining technique and is part of a family of relatively modern material finishing and shaping processes described as "chipless machining". These processes do not use cutting tools and do not create residual surface stresses in the workpiece.

The abrasive flow machining process was developed, during the 1960s, by McCarty[8, 9]. It is a production process for the polishing and surface finish improvement of machined, metallic components. It utilises the controlled extrusion of a semi-solid, abrasive laden, polymer (described as the "medium"). The passage of this pressurised media over component surfaces results in the removal of surface material and, with the use of the correct abrasive, the resultant improvement or polishing of the surface. The advantage of using a semi-solid polymer is that it takes the shape of the area over which it flows and thus acts uniformly over the surface.

The process requires that the workpiece, which must be open in form (i.e. of such a design that it permits the medium to pass through it), is confined within a fixture between two vertically opposed media chambers (or cylinders) within the machine. Under hydraulic pressure, the medium is extruded from one chamber, through the workpiece and into the opposite chamber. This is shown in Figure 1. If required, the operation can be reversed and cycled. Whilst Figure 1 shows a simple straight cylindrical workpiece, the technique can also be applied to workpieces containing cross drillings such as hydraulic manifolds where the use of suitable tooling guides the direction of the media flow.

3 THE ULTRASONIC MACHINING (USM) PROCESS

Ultrasonic machining is also described as a non-traditional machining technique. It is often used in combination with other chipless machining techniques, such as electrodischarge machining, in the manufacture of precision components.

As the name implies, the process operates at ultrasonic frequencies and typical frequencies used in the ultrasonic machining process are 20 and 40 kHz. The equipment used is based on a generator energising a magnetostrictive or piezoceramic transducer which in turn causes an attached tool to vibrate. The vibrating tool transfers energy to abrasive particles carried, within a slurry, underneath the tool tip. The interaction between these particles and the

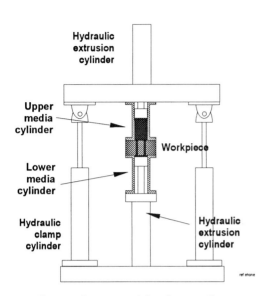

Figure 1 *Diagram of the abrasive flow machining process*

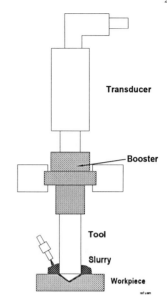

Figure 2 *Diagram of the ultrasonic machining process*

workpiece results in the formation of a mirror image of the tool tip in the surface of the workpiece.

Modern systems tend to use piezoelectric transducers which have an efficiency of 90–95%[10]. They generally use a "booster" which is a stepped cylinder attached to the transducer which is in turn attached to the tool. The change in cross-section of the booster causes amplification of the amplitude of vibration at the tool tip. The tool is often cylindrical in shape but techniques such as finite element analysis and CAD/CAM permit optimisation of the design[11,12].

The transducer/booster/tool assembly is located by a clamp around the midpoint or node of the booster. A typical arrangement is shown in Figure 2.

4 ULTRASONIC FLOW POLISHING (UFP)

An ultrasonically energised abrasive polishing process capable of polishing three dimensional cavities has been developed by an international consortium of academics and industrialists. The process has the capacity to produce a high quality surface finish to the cavity whilst causing minimal deterioration to its profile or dimensional accuracy. Surface finish improvements of up to 10:1 have been recorded and a patent covering the process has been filed[13].

The process combines the two technologies of AFM and USM. The abrasive/polymer mix is pumped down the centre of the ultrasonically energised tool and on exit, its flow constrained by the tool and the workpiece, the mix flows radially relative to the axis of the tool. Whilst constrained, the mix is ultrasonically energised by the vibrating tool. The combination of flow and vibration results in the mix abrading the workpiece surface. Combination with multi-axis

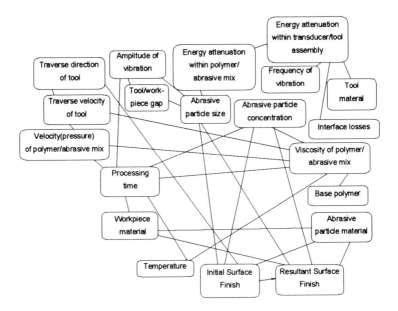

Figure 3 *Variables within the ultrasonic flow polishing process*

CNC tool manipulation allows the polishing of complex three-dimensional dies.

Correct selection of variables such as size, type and concentration of the suspended abrasive particles along with the viscosity of the polymer carrier leads to optimisation of the process.

In a similar manner to both the AFM and USM processes, this ultrasonic flow polishing process is multivariable and stochastic. The variables affecting the process are inter-related and identified in Figure 3.

5 STRESS DISTRIBUTION WITHIN THE ULTRASONICALLY ENERGISED ABRASIVE LADEN POLYMER

In order to characterise the performance of the process, the stress distribution within the ultrasonically energised abrasive laden polymer was determined. Due to the complex rheological properties of these base polymer[14-16], the pressure distribution was determined experimentally and this data used in the development of a pressure distribution model[17].

The causes of system losses were identified as follows:

- Electro-mechanical efficiency of the generator/transducer/tool assembly.
- Transmission losses between the energised tool and the abrasive laden media.
- Coupling losses resulting from poor mechanical connection between the energised tool and the abrasive laden media.
- Attenuation losses within the energised medium as a result of absorption and scattering.

These losses were determined for 20 kHz and 40 kHz SLICE[18] ultrasonic systems.

The model was primarily based upon results measured using the 40 kHz system and it identified that the pressure distribution can be described by:

$$p(z,r) = p(0,0)\big(AF(f,z)\big)(TF)\left\{\frac{R-r}{R}\right\}e^{-0.064z^{1.476}} \quad \{1\}$$

where $p(z,r)$ is the ultrasonic pressure within the polymer/abrasive mix, z is the axial distance from source (the end face of the tool), r is the radial distance from the centre line of the source, $p(0,0)$ is the ultrasonic pressure at the source (which is dependent upon the power setting of the generator and electro-mechanical efficiency of the ultrasonic system) and R is the radius of the tool.

The term $AF(f,z)$ is a dimensionless "Attenuation Factor". It is the ratio of the dynamic pressure within the abrasive laden polymer to the dynamic pressure within the unfilled polymer. Its value is dependent upon frequency and distance from source.

The term TF is also dimensionless and is the "Transmission Ratio". It is based on comparative values of the ultrasonic pressure transmission coefficients determined from the density of the tool or suspension combined with its ultrasonic compression wave velocity i.e.

$$TR = \frac{P_n}{P_a} \quad \{2\}$$

where P is the ultrasonic *pressure transmission coefficient* across an interface. In this case where an aluminium tool is used, P_a is the coefficient across an aluminium/unladen polymer interface and P_n is the coefficient across a non-aluminium/laden polymer interface.

If, on exit from the tool's axial hole, the pressure within the mix is p_f, and r_s is the radius of the axial hole, then the maximum and minimum pressures on the workpiece surface, at a constant gap and power setting, can be described by

$$p_w(r) = \left\{\frac{R-r}{R}\right\}\{p_f \pm 0.5p_u\} \quad (r_s < r < R) \quad \{3\}$$

In the above equation, p_u is the peak-to-peak dynamic pressure within the mix i.e.

$$P_u = p(0,0)\big(AF(f,z)\big)(TF)e^{-0.064z^{1.476}} \quad \{4\}$$

A similar set of expressions can be developed for a 20 kHz, 1 kW system.

6 EXPERIMENTAL METHODS

An experimental test rig was developed in which a 40kHz ultrasonic transducer arrangement was mounted on an adapted CNC milling machine, replacing the standard cutter arrangement. The polymer/abrasive mix was contained within a reservoir and pressurised by means of a simple jack. The pressurised mix flowed along a hose connecting the reservoir to the vibrational node on the tool (approximately mid length). The flow entered the tool by means of a radial

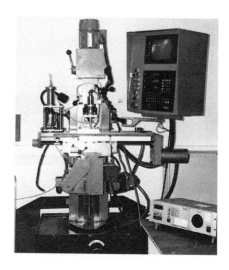

Figure 4 *The ultrasonic flow polishing machine*

hole and exited by an axial hole. On exit it was constrained between the end face of the tool and the milled aluminium workpiece. A photograph of the experimental arrangement is shown in Figure 4.

Commissioning trials of the experimental test rig[17,19] had identified combinations of machining and ultrasonic variables that satisfactorily produced improvement in the surface finish of milled aluminium test pieces. The machining variables related to feed rate, jog, feed direction, tool/workpiece gap, number of passes and the flow inducing pressure within the polymer abrasive mix. The ultrasonic variables related the amplitude of vibration of the tool tip (based upon power setting and tool design).

Maintaining these machining and ultrasonic variables, a programme of ultrasonic flow polishing experiments was carried out, using polymer/abrasive mixes of constant concentration (mass) but containing progressively smaller abrasive particles. The abrasive particles used were boron carbide and the grades were F120, F220, F320, and F600. A similar progressive programme of manual polishing was also carried out using commercial abrasive cloths. The cloth grades were F150, F220, F400, F600 and F800.

In the case of the ultrasonic flow polishing experiments, the polishing was carried out for 3 cycles. For the manual polishing experiments, the polishing with a particular abrasive size was continued until there was no further, visible improvement in surface finish.

7 RESULTS

The results achieved by the two different polishing processes are presented in the form of micro-photographs in Figures 5 and 6 and in the form of surface profiles in Figures 7 and 8.

8 DISCUSSION

It is acknowledged that there is an "art" to the polishing of aluminium, and the manually

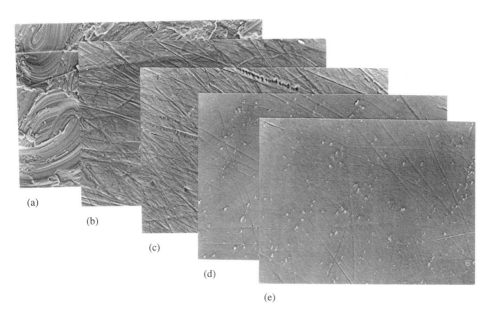

Figure 5 *Micro-photographs of progressively, ultrasonically polished aluminium: (a) as machined, (b) post F120, (c) post F220, (d) post F320 and (e) post F600*

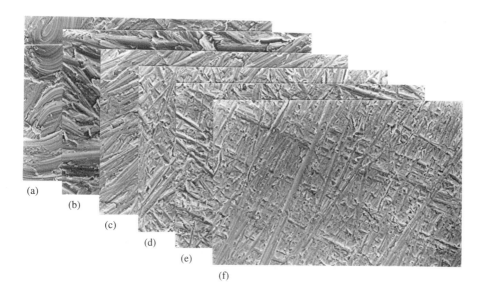

Figure 6 *Micro-photographs of progressively, manually polished aluminium: (a) as machined, (b) post F150, (c) post F220, (d) post F400, (e) post F600 and (f) post F800*

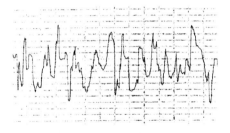

a) As machined - Ra 1.86, Ry 12.3, Rz 10.3

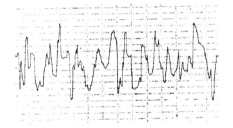

a) As machined - Ra 1.86, Ry 12.3, Rz 10.3

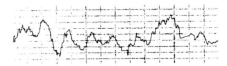

b) Post F120 - Ra 0.91, Ry 6.7, Rz 4.67

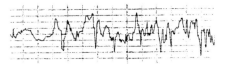

b) Post F150 - Ra 1.12, Ry 8.43, Rz 7.06

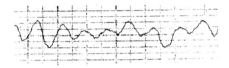

c) Post F220 - Ra 0.75, Ry 4.03, Rz 3.40

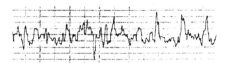

c) Post F220 - Ra 0.78, Ry 6.63, Rz 5.63

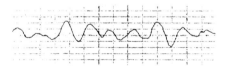

d) Post F320 - Ra 0.63, Ry 3.9, Rz 2.70

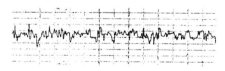

d) Post F400 - Ra 0.46, Ry 4.07, Rz 3.40

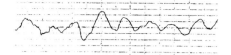

e) Post F600 - Ra 0.72, Ry 4.4, Rz 3.13

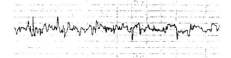

e) Post F600 - Ra 0.47, Ry 4.23, Rz 3.47

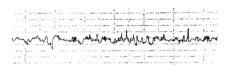

f) Post F800 - Ra 0.37, Ry 3.07 Rz 2.67

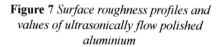

Figure 7 *Surface roughness profiles and values of ultrasonically flow polished aluminium*

Figure 8 *Surface roughness profiles and values of manually polished aluminium*

polished samples were not polished by experienced personnel. This was deliberate and intended to simplify initial comparison between manually and ultrasonically polished surfaces.

It would appear that in the case of the manually polished samples all the original machining marks were removed as a consequence of the abrasive action of the F150 grit. This resulted in a relatively rough but homogeneous surface finish. This homogeneous finish was progressively improved as the process continued, and smaller and smaller abrasive particles were used. The end result was a finely scratched surface.

In terms of roughness values, the action of the ultrasonically energised F120 abrasive particles resulted in a greater improvement in surface finish than the equivalent (F150) manual process. However, the results suggested that the initial surface finish was not homogenised. As a consequence, the subsequent polishing steps, whilst they locally improved the surface finish, failed to significantly improve the overall surface finish to the same degree as that obtained by the manual process.

It is proposed that there may be two causes for the poor surface after the initial pass. Firstly, the manual polishing of aluminium results not only in the scratching of the surface but also the smearing of the surface. It is possible that the forces and temperatures generated in the ultrasonic process were not sufficient to cause smearing. Secondly, the flow of the polymer/abrasive mix followed the least constricted path across the tool/workpiece interface and, as a consequence tended, to undercut rather than remove the highspots.

It has been identified earlier that within the ultrasonic process there is considerable interaction between the variables. As a consequence, each abrasive/polymer mix will have its own set of optimised variables. At present, these have not been fully determined.

9 CONCLUSIONS

An empirical model of the stress distribution within ultrasonically energised polymer/abrasive mixes has been developed. The use of the model as a method of characterisation of the process will permit control and optimisation.

Results have been presented which demonstrate the successful development of an ultrasonic flow polishing process. The results were achieved using a 0.25 kW, 40 kHz ultrasonic system. The introduction of a more powerful 20 kHz system and further work to determine the optimum matching of the variables within the process has the potential to produce a more homogenous surface to the workpiece. This homogeneous surface will result in significant improvement in the final polished surface.

Acknowlegements

This research was conducted with the assistance of the CEC, (BRITE-EURAM Project No BE-4518(89)).

References

1. K. Saito, 'Automatic grinding and polishing of curved surfaces'. Press Technol., 1983, 21, p. 18, (in Japanese).

2. L. J. Rhoades, *J. Mat. Processing Technology*, 1991, **28**, 107.
3. T. A. Kohut, Surface finishing with abrasive flow machining, Aluminum Extrusion Technology 4th Int Seminar, Chicago, Ill, USA, 1988, Vol. 2, p. 35.
4. R. E. Williams, Metal removal and surface finish characteristics in abrasive flow machining. MSc Thesis, University of Nebraska, 1989.
5. V. Soundararajan and V. Rahakrishnan, *Int. J. Mach. Tool Des. Res.*, 1986., **26**, 307.
6. A. B. E. Khairy, *Wear*, 1990, **37**, 187.
7. D. Kremer and J. Mackie, *L'Industrie Ceramique*, 1988, **830**, 632.
8. E. Hoffman and R. McCarty, Machine with Putty?....Silly Putty?? Tool Prod, Feb. 1966, p. 54.
9. R. W. McCarty, Method of honing by extruding. US Patent No 3 521 412, 1970.
10. J. A. McGeogh, Advanced methods of machining, Chapman and Hall, London, 1988.
11. G. Coffignal, M. Touratier and M. Frainais, 5th Int Modal Analysis Conf, London, England, 1987, p. 1438.
12. S. G. Amin, M. H. M Ahmed. and H. A. Youssef, AMPT'93, Dublin, Eire, 1993, p. 1455.
13. Extrude Hone Ltd., Method and apparatus for ultrasonic working. UK Patent Application No 94 19749.8, 1994.
14. J. B. Hull, A. R. W. Heppel and A. R. Jones, Surf Eng, 1992, **2**, 234.
15. L. J. Rhoades, 'Modelling the flow behaviour of industrial borosiloxane', NSF-SBIR Phase 1 Final Report, 1985.
16. S. A. Trengove, PhD Thesis, Sheffield Hallam University, 1993.
17. A. R. Jones, Unpublished Communication, PhD Thesis, submitted to University of Bradford, 1995.
18. B. Thirion and P. Poupaert, 'Procede et dispositif d'alimentation electrique d'un transducteur generateur de vibrations tant sonores qu'ultrasonores'. French Patent No 2586883, 1985.
19. A. R. Jones and J. B. Hull, AMPT'95, Dublin, Eire, 1995, p. 867.

Section 2.5 Electrochemical and Electroless

2.5.1

The Influence of Formic Acid and Methanol on Deposition the Hard Chromium From Trivalent Chromium Electrolytes

S. K. Ibrahim[1], A. Watson[1] and D. T. Gawne[2]

[1]DEPARTMENT OF CHEMISTRY AND CHEMICAL ENGINEERING, UNIVERSITY OF PAISLEY, SCOTLAND.

[2]SCHOOL OF ENGINEERING SYSTEM AND DESIGN, SOUTH BANK UNIVERSITY, LONDON, ENGLAND.

1 INTRODUCTION

Metallurgists have always been interested in the engineering properties of chromium since it provides excellent wear and corrosion resistance. These coatings are normally obtained from hexavalent chromium electrolytes, which are unfortunately not greatly desirable because of disadvantages such as toxicity and poor throwing power etc.

Therefore, extensive work[1-5] has been conducted over many years to develop chromium electrolytes incorporating trivalent chromium to replace the more toxic chromium (VI) process. However, for many years the most successful attempts could produce high quality deposits only for short deposition times and the thickness obtained was suitable for decorative applications only. In the last few years thick coatings have become possible but with low deposition rates typically falling to about 1μm/hr on prolonged electrolysis. Exceptionally, a very recent study[6] has claimed that a deposition rate of 300 μm/hr was obtained using special conditions such as high flow rate and high current density.

In this work, before studying the engineering properties of the deposit from trivalent chromium electrolytes, it was decided to investigate the possibility of developing a chromium (III) electrolyte to provide high deposition rate with good quality deposits using conventional conditions.

The influence of formic acid, methanol and their combination on the electrodeposition of chromium (III) process was examined. The effect of methanol in particular was studied in detail since it allowed a great increase in the deposition rate and the quality of the deposit. Moreover, a similar effect of methanol had been already reported[7], where its addition to chromium (VI) baths had increased the current efficiency and the quality of the deposit. Benaben[8] also reported an electrolyte formed by addition of methanol to chromium (VI). Both of these electrolytes are in practice chromium (III) electrolytes.

The change in the electrolyte chemistry and electrochemistry was also investigated, which was believed to play an important role in controlling the deposition rate and the quality of the deposit. An ion-exchange chromatographic method was successfully used to identify the various species of chromium for during the complexation reaction with ageing time. The concentration of chromium in each fraction (band) was determined by ICP spectroscopy. Identification of the fractions was also further confirmed by UV visible spectroscopy.

Excellent quality and high plating rate were obtained by the addition of formic acid and methanol. The deposition rate was enhanced up to an optimum of 50 μm/hr and was easily raised up to 100 μm/hr by employing slightly higher current density. A sustained deposit up to 200 μm was also obtained with prolonged electrolysis. Further, conventional conditions such as 25°C, low current density and a gentle agitation were used.

2 EXPERIMENTAL DETAILS

2.1 Electrolyte

The electrolyte was prepared to the following composition:

0.8M	$[Cr(H_2O)_4Cl_2]Cl2H_2O$
0.5M	NaCl
0.5M	NH_4Cl
0.15M	$B(OH)_3$
0.2M	HCOOH
2.0M	H_2NCONH_2
200ml	CH_3OH

All the salts were first dissolved together in UHP water, except for the chromium salt which was dissolved separately. Then formic acid and methanol were added, and the dissolved chromium salt added last. Finally, the solution was made up to the mark of 1 dm^3 in a volumetric flask. The electrolyte was employed at make up pH without further adjustment, unless otherwise indicated, where HCl and NaOH were used.

2.2 Electrodes

The working electrodes (cathode) were prepared from high conductivity AnalaR grade copper sheet (Goodfellow Metals Ltd.). The copper sheet was cut into 5 x 2.5 cm^2 pieces, these were degreased twice by soaking in acetone for a few minutes each time. The degreased samples were coated with non-conducting Lacquer to leave an exposed area of 2.5 x 2.5 cm^2.

Prior to use, the copper cathodes were anodically electropolished at 15V for 30 second and then at 0.9V for 3 minutes, followed by rinsing with flowing de-ionised water then transferred wet to the plating cell. The electropolishing solution was used consisted of 90% v/v phosphoric acid, 5% v/v water and 5% v/v methanol.

Insoluble high density graphite was used as the anode material. The reference electrode was a standard calomel electrode (SCE).

2.3 Plating System

Figure 1 shows a schematic diagram of the electrodeposition apparatus. A rectangular thin walled plastic cell (10 cm length, 4 cm width and 8 cm depth) was used. A gentle agitation of the electrolyte was used by circulating it between the cell and a 2 litre reservoir. For this purpose the cell was fitted with an outlet leading via a peristaltic pump (Watson and Marlowe) to the reservoir, with return by gravity feed to the cell. The flow was from anode to cathode, with a speed of 200 ml/min. The temperature of the electrolyte was maintained at 25 ±2°C.

A constant potential was imposed by using a potentiostat. A ministat 253 (Sycopel Scientific

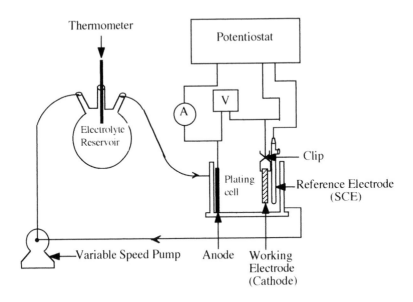

Figure 1 *A scematic diagram of the electrodeposition apparatus*

Ltd) was used. Both the cell potential and resulting current were measured using Iso-tec IDM91 digital multimeters. A constant current supplier (Farrnell L30-2) was also used in the electropolishing process.

2.4 Quality and Rate of Deposition

The quality of the metal deposits was assessed visually, and using optical microscopy and electron microscopy in terms of their porosity, structure, pit cracks and microcracks. Adhesion was tested by bending through 180° and examine under low power microscopy. The other physical properties of the deposit such as ductility, hardness and wear resistance will be the subject of forthcoming publication. High quality deposit was defined by white metallic deposit of good appearance and good adhesion with no pitting or porosity.

Deposition rates were determined from thickness versus deposition time. This involved two methods. The first was by measuring the change in weight of the sample after deposition and calculating the thickness of the deposit, from the weight of deposit, the density and the plated area. In the second, the thickness of the deposit was obtained directly by the cross sectioning of the sample and examining it using Buehler Omnimet-l imaoe analysis system.

2.5 The Ion Exchange Chromatographic Technique

Ligand exchange reactions involving chromium (III) ions occur very slowly, and as a result, with electrolytes prepared at room temperature, changes in the coordination of the

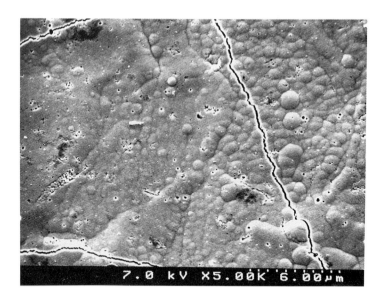

Figure 2 *SEM photograph of a chromium deprosit from the urea, formic acid and methanol based electrolyte at −2.4 V (SCE) and 200 ml/min flow rate*

chromium(III) ions take place throughout the lifetime of the electrolytes. These changes in the complexation which may be responsible for the deterioration of the quality of the deposits and the deposition rate have not been studied in detail before. In this study, ion exchange chromatographic studies of chromium (III) base electrolyte were used to provide some useful information on the changes in the chemistry of the bath during prolonged electrolysis, relating to the reduction of either the quality or the deposition rate.

A typical sample of 8 ml of a solution with total chromium concentration of 431.0 ppm was taken originally from a solution of 2 ml of the working electrolyte diluted to 25 ml in 0.1M $NaClO_4$. The pH of the solution was adjusted with $HClO_4$ to a value of 1.80±2. The 8 ml sample was adsorbed on the cationic ion-exchange resin (Sephadex SPC-25) in a 8 × 1 cm column. The elution procedure employed was similar to that reported by Stünzi and Marty[9]. Various species (bands) depending on the age of the electrolyte, as discussed later, were observed. The bands were collected and appropriately diluted in volumetric flasks for quantitative analysis of total chromium in each band. A Perkin Elmer optima 3000 ICP-AES system was used. The bands were also further identified spectrophotometrically using a Perkin Elmer Lambda 9 UV/VIS/NIR spectrophotometer.

3 RESULTS AND DISCUSSION

Initial work showed that addition of carefully controlled quantities of methanol and formic acid, greatly enhanced both the deposition rate and quality obtained from electrolytes based on both dimethylformamide (DMF) and urea. These chloride electrolytes had previously been

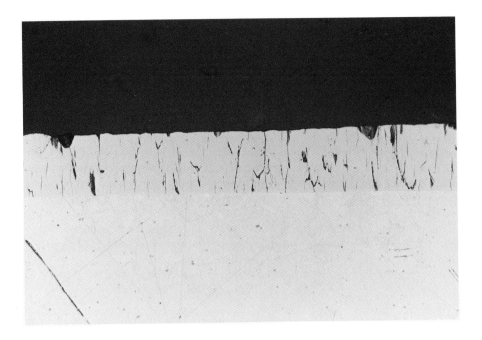

Figure 3 *Cross-section of a chromium coating deposited at » mm/hr from the urea, formic acid and methanol based electrolyte at –2.4 V (SCE) (× 400)*

investigated without such addition[2], mostly for chromium alloy deposition.

In the current study urea was used as the main complexing agent rather DMF, which has been commonly employed in this type of bath, since it is not toxic and commercially available. It is also very soluble in water. Urea has been used in large quantities[10] and as the major constituent in many electrolytes[11]. However, the initial quality of the deposit front the original urea electrolyte with chromium (III) as the only metal could be described only as fair. It improved dramatically on the addition of formic acid and methanol as discussed below.

As in any chemical process, there are various variables which may affect performance. Therefore, it is necessary to examine the influence of varying conventional parameters. Various chromium (III) salts as the base electrolyte were also investigated.

3.1 A Study of Various Chromium (III) sources

A comparison of three chromium (III) electrolytes systems containing, $CrCl_3 6H_2O$, $CrK(SO_4)_2 12H_2O$ (chrome alum) – $Cr_2(OH)(SO_4)_2 Na_2 SO_4 10H_2O$ (chrometan) or $Cr(ClO_4)_3 6H_2O$ has been carried out. In each case the other salts were also changed from chloride to sulphate or perchlorate as appropriate. The best quality and highest plating rate were obtained from chloride electrolytes with the inclusion of formic acid and methanol as shown in Figures 2 and 3 respectively. The pH of the electrolyte at make up was ~ 1.8 which dropped down to ~ 0.6 after 15–20 minutes of electrolysis. This period was found to be sufficient for the electrolyte to reach stability, particularly pH, to produce an excellent quality of deposit with high plating rate up to ~ 50–100 μm/hr.

Two types of sulphate electrolytes, a violet modification (chrome alum), hexaaquo chromium (III), and a green modification (chrometan), a sulphato complex, were used.

On exchanging the chloride salt for 0.1 M of the first of the sulphate salts (chrome alum), the electrolyte did not perform very well. The deposit was either incomplete at the centre of the cathode surface or spoiled by a black non-metallic deposit around the outside edges. Addition of formic acid, methanol or both together had no effect on the performance of the electrolyte or the deposition rate.

Replacing chloride salts by the second sulphate salt (chrometan), the sulphato complex, the electrolyte behaved better than the chrome alum. The general quality of the deposit was slightly better, but more importantly, complete coverage was obtained. However, the deposit was still grey and dull, going to a very dark grey-black colour on prolonged deposition with maximum plating rate of 1.0 μm/hr. Addition of methanol and formic acid had no effect on the deposition process or rate.

It was also found that the methanol did not affect the deposition rate from a chrometan/malate/sulphate based electrolyte, where an excellent quality of deposit was obtained with and without it, but at very low and falling deposition rates.

It is likely that the sulphate electrolytes, encourage the formation of stable sulphato bridged oligomers, which hinder the deposition process, and are little affected by the addition of methanol or formic acid.

Chromium perchlorate was also studied as an alternative to chloride to further reveal the rôle of the chloride ions in the electrodeposition process, since neither can form bridged species in aqueous solution. A black non-metallic deposit was obtained, although the make-up pH of the electrolyte ~ 1.40, was lower than that of the chloride electrolyte pH ~ −1.80. The presence of methanol and formic acid had no effect.

It thus appears that the presence of chloride ions is a very important factor in the electrodeposition from this chromium (III) electrolyte, which is believed to be due to its catalytic effect on the electron transfer process, and greater ease in its replacement by other ligands.

It has also been shown from the chromatographic study (as discussed later) that the formation of higher molecular weight species as below, was delayed in the presence of chloride ions while they appeared within ~ 4 hrs of electrolysis in the presence of perchlorate or sulphate ions. Sulphate salts are still under further investigation in the meantime, and will be reported in a further publication. However, it has been reported[12] that olation reactions as shown below are more accessible in the presence of sulphate ions, which can participate in the bridging leading to especially stable oligomeric forms and thus interfere with the deposition process and lead to reduced deposition rates and eventually total deposition failure.

Therefore, it was decided to concentrate on the investigation of the chloride electrolyte and some of the factors which may affect the deposition process.

3.2 The Effect of Formic Acid

Formic acid has been regularly included as a part of obtaining a successful deposition process from Cr(III) electrolytes[13]. Its general rôle unfortunately, is complex and not very clear, although, it's one of vital components in the current process. It has been used in the past[12] at various concentration levels and appeared to have behaved in quite different ways in different electrolytes – indicating a variety of different mechanisms open to it. For example, Anderson[14] has attributed the importance of the formic acid presence in controlling the olation reaction to controlling the Cr(II) level in the diffusion layer.

However, in this study the possibility was considered for investigation that formic acid could act in numerous ways:

1. A catalyst to chromium(III) ligand exchange.
2. By forming the quadruply bonded dimeric chromium(II) the formic acid would reduce the ability of transient chromium(II) to catalyse olation reactions in the diffusion layer.
3. As a complexing agent for chromium(III), possibly even a bridging ligand at higher formic acid concentrations.

These possibilities were investigated by a detailed study of the pH characteristics of the electrolyte, including the cathode diffusion layer, with and without the formic acid, and by speciation of these electrolytes by ion exchange chromatography associated with ICP spectroscopy and UV/VIS spectroscopy.

However, initially it was shown that lowest levels of formic acid are less effective in accelerating the deposition rate, while higher levels lead to a major loss of deposition quality and to the failure of the electrolyte. This indicates the complex role of formic acid in these electrolytes. The optimum concentration of formic acid (given in the Experimental section,) is much lower than the chromium, consistent with the positive role being either a catalytic role or as a complexing agent with an electrolysis intermediate, while the negative role at higher concentrations may involve relatively inert chromium(III) complexes, possibly bridged. This will be further discussed in later section.

3.3 The Effect of Methanol and its Concentration

To the authors' knowledge the effect of methanol on existing chromium (III) electrolytes has not been studied before. However, a UK patent was published in 1985[7] claimed that adding methanol to chromium (VI) bath increased the current efficiency and quality of chromium-iron alloy plating. The patent has ascribed the effect of methanol as due to the following reaction:

$$4CrO_3 + 3CH_3OH + 12H^+ \longrightarrow 3HCOOH + 4Cr^{+++} + 9H_2O$$

With this reaction occurring, formic acid is produced. In practice there will be considerable further reaction to carbon dioxide and water. As the methanol is in excess their electrolyte will contain methanol and some formic acid. The authors of this patent regarded the reaction as a source of chromium(III), they noted the presence of formic acid but did not investigate its role. Benaben[8] also reduced chromium (VI) with methanol but ignored the likely organic products. In this study, formic acid and methanol have been clearly demonstrated to accelerate the deposition rate and quality of existing well developed chromium(III) electrolytes based on dimethylformamide and urea. The dimethylformamide electrolytes behaved in a very similar way to the urea electrolytes reported here.

The authors believe that the reduced deposition rates for chromium (III) electrolytes are due to adsorption of olated oligomers of chromium(III), that is μ-hydroxo bridged complexes of chromium(III) at the growth site on the deposit surface. These olation reactions are favoured by increased pH and by catalysis by chromium(III) of the slow ligand exchange reactions of chromium(III).

The following possible roles for the methanol, in removing this hindrance to deposition, had to be considered and investigated:

1. The methanol could lower the pH of the cathode diffusion layer, by reducing hydrogen evolution. This would hinder olation reactions.
2. The methanol could assist in desorbing surface active μ-hydroxochromium(III) oligomers.
3. The methanol could form complexes with the Chromium(III).
4. The methanol could act in the salvation sphere of the chromium(III) to inhibit aggregation, as it does with dye stuffs at these concentration levels, and hence to inhibit olation.

Before investigating these possibilities by a detailed study of the pH characteristics of the electrolyte in similar manner of the one mentioned in the effect of formic acid section, it is relevant to study both of the effect of various concentration of methanol and the use of different alcohols such as ethanol, etc.

Table I shows the effect of various concentration of methanol on the quality of the deposit. It appeared that both lower and high concentration did not affect the quality dramatically.

In contrast the deposition rate was found to be affected significantly by the concentration

Table 1 *The effect of methanol on the quality of deposit*

Vol. (ml/l)	Quality
50	Good/bright
100	Good/bright
150	Excellent/bright
200	Excellent/bright
300	Very good/bright

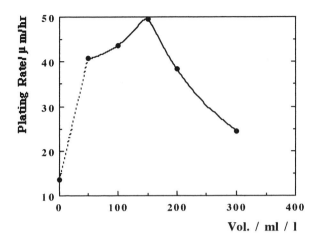

Figure 4 *The effect of various concentration of methanol on deposition rate*

as shown in Figure 4. Relatively small quantities of methanol produced a marked increase in deposition rate rising to a maximum about 20%. Beyond this the deposition rate falls rapidly. The latter is associated with loss of solubility and decrease in conductivity.

Further investigation was carried out by studying various solvents as following:
Ethanol: Table 2 shows the effect of ethanol on both the quality and deposition rate. A deposition rate up to ~ 38 μm/hr was obtained, Figure 5, which is lower than in methanol by ~ 33%. However, the quality of the deposit was almost similar to those obtained with methanol as shown in Figure 6.
t~Butanol: This also yielded a good deposition rate as shown in Figure 7, similar to ethanol as shown in Tables 2 and 3, as well as a good quality of deposit as shown in Figure 8. However, the volume of this solvent used to achieve the optimum effect was lower than ethanol and methanol which may be due to the effect of the viscosity, which is much higher compared to the others.
Acetone: This solvent showed much lower plating rate than the others, where a deposition rate of ~ 10 μm/hr was obtained with a little decline in the quality of the deposit.

Table 2 *The effect of ethanol on both deposition rate and quality*

Vol (ml)	Deposition Rate (mm/hr)	Quality
50	36.91	Very good/bright
100	35.86	Good/dull/black edges
200	-	It does not work

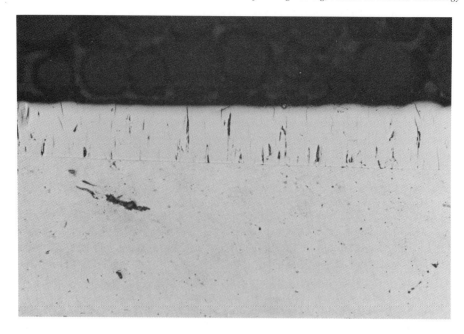

Figure 5 *Cross section of a chromium coating deposited from the urea, formic acid and ethanol based electrolyte at –2.2 V (SCE) (×400)*

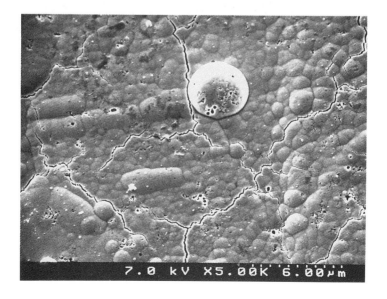

Figure 6 *SEM photograph of a chromium deposit from the urea, formic acid and ethanol based electrolyte at –2.2 V (SCE) and 200 ml/min of flow rate*

Table 3 *The effect of t-butanol on both deposition rate and quality*

Vol (ml)	Deposition Rate (mm/hr)	Quality
50	37.78	Fair/bright with grey lines
100	36.90	Fair/dull/black edges
200	35.06	Very good/bright

The effect of the above organic solvents on the properties of electrolyte solutions is rather complicated since it depends on several solvent characteristics themselves such as, polarity, the dipole moment, the dielectric constant and the electron pair and/or hydrogen bond acceptance and donation abilities etc. Table 4 shows the physical properties of solvents.

For example the above results showed that the order of deposition rate with methanol, ethanol and t-butanol are consistent with the order of the dielectric constants, which are 32.63, 24.30 and 10.9 respectively. However, the later one is less than half that of the ethanol, but they both showed quite similar results regarding the quality and the deposition rate. This would indicate that the electrolyte properties do not simply follow the order of the dielectric constant alone, where other factors also need to be considered.

The situation with acetone is anomalous. According to the dielectric constant, it would be expected that acetone should give a deposition rate above that obtained with t-butanol, as its

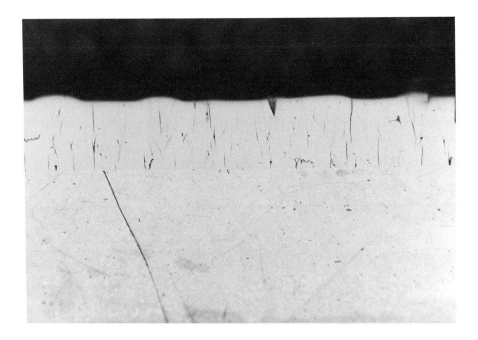

Figure 7 *Cross section of a chromium coating deposited from the urea, formic acid and t-butanol based electrolyte at –2.2 V (SCE) (×400)*

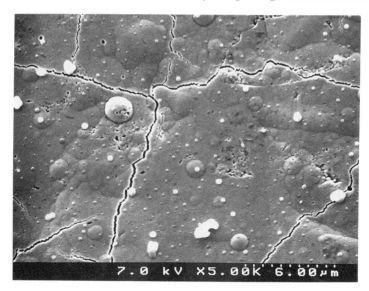

Figure 8 *SEM photograph of a chromium deposit from the urea, formic acid and t-butanol based electrolyte at –2.2 V (SCE) and 200 ml/min of flow rate*

dielectric constant is 20.7. In fact acetone showed the lowest plating rate(~10 μm/hr) among these solvents, although, similar quality was obtained. That would again indicate the effect of solvent is not simply controlled by one parameter.

It is believed that these solvents enhance the deposition rate by acting through the salvation sphere of the chromium(III) species to inhibit the aggregation necessary to the olation reactions which produce the higher molecular weight oligomeric species which inhibit the deposition process, through adsorption on the metallic growth site. A similar inhibition of aggregation by the solvent acting through the solvation sphere has been reported for dye stuffs at similar solvent concentrations levels[15]. Such an inhibition of aggregation would in part be dependent on the dielectric constant.

Table 4 *Physical Properties of solvents at 25°C*

Solvent	Dielectric Const (ε)	Polarity	Viscosity (cp)	Cohesive energy density $(cal/cm^3)^{1/2}$	Plating rate (μm/hr)
Methanol	32.63	0.39	0.547	14.5	49.79
Ethanol	24.30	0.27	1.200	12.7	37.78
t-Butanol	10.90	0.11	5.942	10.4	36.91
Acetone	20.70	0.69	0.303	10.0	10.00

There is likely to be a relationship between the stability of the solvent in the salvation sphere and the cohesion of the solvent, which in turn would affect the ability to inhibit aggregation. This cohesive attraction can be approximated as the cohesive energy density given as the energy of vaporisation per molar volume. This is often given as its square root and is also known as the solubility parameter. In this form it is often used to judge the compatibility of solvents. In a mixed organic/aqueous solvent system the cohesive energy density is likely to have a role in the solvent structure and in the structure and composition of the salvation sphere. The enhancement of the deposition rate in the presence of methanol, ethanol, t-butanol and acetone is in the order of the cohesive energy densities, 14.5, 12.7, 10.4 and 10.0 $(Cal/cm^3)^{1/2}$ respectively. Acetone, although in the correct position in the order, offers only an anomalously low enhancement to deposition rate.

A further factor is the donor power of the solvent, the ability to enter into the inner coordination sphere of the chromium atom. Alcohols have little tendency to coordinate to the metal. The ion exchange chromatography showed no sign of such complexes. In contrast, acetone is likely to have a limited donor power through its ketone group. The appearance of new bands in the ion exchange chromatography with time confirms this. Acetone yields only a very limited enhancement of the chromium deposition rate. Dimethylformamide and triethanolamine are stronger complexing agents; they produced no enhancement of the deposition rate, although their dielectric constants and cohesive energy densities are not dissimilar to the alcohols.

We can thus conclude that for the addition of a solvent to enhance the chromium deposition rate, the solvent should be a strongly polar solvent to dissolve the salts, have as high as possible a dielectric constant and cohesive energy density and NO tendency to form complexes directly with the chromium. Methanol thus appears to be the best solvent for this purpose.

3.4 pH and Electrodeposition

3.4.1 Changes in pH on Electrolysis. These electrolytes have a pH about 1.6–1.8 on make up. Figure 9 shows that the complete electrolyte with formic acid and methanol undergoes

Table 5 *pH of the cathodic diffusion layer as a function of deposition time. Bulk pH 1.80, Cathode Potential –2.4 V (SEC) and Temperature 25°C*

Deposition time	Diffusion layer pH		
	+CH₃OH/HCOOH	+HCOOH	Neither
0.5	3.47	3.70	3.73
1.0	3.52	3.76	3.78
3.0	3.53	3.77	3.80
5.0	3.56	3.80	3.83
10.0	3.52	3.86	3.85

a rapid fall in pH to 0.5–0.6 in about 15–20 minutes of electrolysis, at which point a steady value is maintained. It appears to take a little longer but less than an hour to achieve initial complete complexation with the urea. This period can be referred to as pre-ageing. The electrolyte then shows no major changes until some 160 hours (for 1 litre of solution at 25°C at –2.4 V) have passed.

Without formic acid and methanol the pH changes are much less rapid or so extreme. For an electrolyte without formic acid or methanol the pH falls on electrolysis to pH ~ 0.98 and pH ~ 1.7 for urea concentration of 2.0 M and 8.0 M respectively. The former urea concentration allows only thin deposits of high quality, while the latter allows thick layers to be deposited of fair quality. The period of pre-ageing to initial complete complexation is 2–3 hours of electrolysis in the absence of formic acid and methanol.

It was decided to adjust the electrolytes without methanol or formic acid to pH 0.5, to see if this offered an improvement in deposition rate or quality. For both 2.0 M and 8.0 M urea metallic deposition failed with black non metallic deposits. This indicates that the improved deposition rates and quality on addition of methanol and formic acid are not a simple matter of the reduced working pH.

Addition of formic acid but not methanol allows a significant improvement in quality and deposition rate. However although thick deposits become possible at 2.0 M urea on addition of formic acid alone, the thickness is limited to maintain adequate quality. The best quality thick deposits, with maximum deposition rate were obtained when both methanol and formic acid were added.

3.5 The Diffusion Layer

The study of the pH of the diffusion layer using the Brenner drainage technique[16] and micro pH electrode showed that methanol has no significant direct effect on the cathode diffusion layer pH. A comparison between the electrolyte with and without methanol and

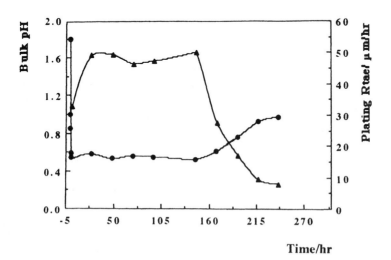

Figure 9 *The effect of electrolysis time on the bulk pH (•) and the rate of deposition (◆)*

formic acid was carried out as shown in Table 5. The bulk pH of these electrolytes was adjusted to the same value for comparison.

As the bulk pH of the three electrolytes is lowered towards pH 1 there is only a minor effect on the diffusion layer pH. Below this bulk pH value the diffusion layer pH drops rapidly towards 2.0–2.5, but for the same low bulk pH the diffusion layer pH values of the three electrolytes remain about equal, with the methanol/formic acid containing electrolyte only about 0.2–0.3 units more acid.

These results show clearly that the reduced hydrogen evolution, on addition of methanol does not directly lower the cathode diffusion layer pH significantly, and so would itself not have a direct effect on olation. The reduction of gas evolution also would not result in the rapid drop in bulk pH with electrolysis on the addition of methanol. This clearly points to a change in the speciation of the electrolyte as was in fact found on ion exchange chromatography.

3.6 The Relationship of pH and Deposition Time From the Methanolic Electrolyte

Within 15–20 minutes of deposition from the electrolyte with addition of methanol and formic acid, the pH drops to a value of pH 0.5–0.6 and remains at this value for a considerable period of rapid reproducible deposition (50–100 µm/hr) of excellent metallic quality. Indeed, the study of the change in pH over a period of 240 hrs showed the pH was quite steady for time to $\approx$ 168 hrs as shown in Figure 9.

Although no dramatic change in the pH up to 168 hours was observed, a gradual loss, initially slight, of the quality of the deposit was found after only @ 100 hours of electrolysis, which was ascribed to minor changes of the chemistry of the electrolyte. However, the drop of the deposition rate has not started until after ~ 144–168 hrs, (Figure 9). The ion exchange chromatography clearly showed that higher molecular weight species started forming at that period of time, as will be discussed later. The loss of deposition rate is believed to be due to the adsorption of the high molecular species on the cathode surface, especially at the growth points.

Figure 9 shows that a deposition rate of 30 µm/hr was initially obtained from a fresh electrolyte due to incomplete complexation. After reaching the optimum, the deposition rate, (50 µm/hr) remained constant for a long period until 144-168 hrs. It started falling down to 8.0   µm/hr after 240 hrs of electrolysis. It was also found that the bulk pH started rising at the same time. The ion exchange chromatography has shown that this sudden rise in pH is associated with the loss of a particular species from the electrolyte which is presumably highly acidic. This contributes to the loss of quality and deposition rate, as the rise in pH would lead to extensive oligomerisation. The appearance of such oligomeric species was first detected by the ion exchange chromatography at this time in the electrolysis. It is believed that the adsorption of these oligomers on the crystal growth site of the metal leads to both loss of deposition rate and a lowering of metallic quality.

In the absence of formic acid and methanol the preageing period to complete complexation is longer, the low working pH is not achieved (pH >1) and the collapse of the deposition rate occurs in less than ~ 72 hours. The oligomeric species are detected very early in the electrolyte lifetime. Thus in addition to yielding an accelerated deposition rate and higher metallic quality, the addition of both formic acid and methanol markedly increase the stability of the electrolyte. The addition of formic acid alone significantly improves the deposition rate, however, it has only a minor effect on delaying the final collapse of the deposition rate. It is the methanol which prolongs the optimum working lifetime of the electrolyte.

3.7 Electrochemical Ageing and Chromatographic Studies

An electrochemical ageing study was carried out on the electrolytes for up to 240 hours of electrolysis. Samples were taken at regular intervals and subjected to ion exchange chromatography. The separated bands were characterised by UV-visible spectroscopy, and their chromium content determined by ICP-AES spectroscopy. The results are summarised in Tables 6 and 7. Likely assignments of the species with spectroscopic data from the literature are given in Table 8.

The ion exchange chromatographic process separates the chromium ions according to charge and approximate size. The larger positive charges and molecular weights species are more firmly held on the column. The information given is as to whether the chromium(III) species is a monomer dimer trimer etc. and indicating its charge. No negative or neutral species were found in this electrolyte.

Fresh electrolytes were found to produce two bands. The first, band 0 was a dark yellow green colour and the second, band 1, a paler bluish green. The position of both bands in the elution sequence correspond to monomeric species with a charge of 1+ and 2+ respectively, indicating a dichloro and monochloro species respectively. UV-visible spectroscopy gave λ_{max} values consistent with $[Cr(H_2O)_4Cl_2]^+$ for the first yellow green band. A shift in the λ_{max} indicated that urea has replaced some water ligands in the monochloro complex, band 1.

In the presence of formic acid band 0 the dichloro form disappears rapidly within the first hour of electrolysis. Without formic acid or methanol it disappears much more slowly taking about 3 hours of electrolysis. On its own methanol has little effect on the removal of the dichloro form. This process which can be followed by the human eye, as the electrolyte loses its yellowish tinge, corresponds to pre-ageing, or completion of complexation. In this process the formic acid appears to act as a catalyst to ligand exchange as urea completes its complexation with the original dichlorochromium(III) source salt.

After I hour of electrolysis traces of a new band, band 2, appear. After 24 hours this has become a major band. Careful observation shows that this consists of two closely overlapping bands, 2b and 2g, The slightly faster moving band 2b is purplish blue while the slightly slower is a greenish blue. The position of the bands corresponds to triply charged monomeric forms. The UV-visible spectra of the band 2b has λ_{max}, corresponding to the simple hexaaquo ion $[Cr(H_2O)_6]^{3+}$. Band 2g could not be isolated and the mixture of the two bands had λ_{max} of higher wavelength than the hexaaquo, consistent with a hexaaquo complex such as $[Cr(urea)_3(H_2O)_3]^{3+}$. Such a species would be eluted from the chromatographic column slightly behind the hexaaquo form.

The greenish-blue band 2g appears to be critical to the accelerated deposition. It remains present in the electrolyte with formic acid and methanol, until some 140–160 hours of electrolysis. Its disappearance after 168 hours corresponds to both the collapse in the deposition rate, and the rise in the bulk pH. The species involved is believed to be very acidic and most likely responsible for the sharp drop in bulk pH on the start of electrolysis.

A third very weak band, band 3, appears slightly later than band 2. Its position in the elution sequence and the associated UV visible λ_{max} indicates it to be a hydroxobridged dimer. The band remains small and this species does not appear to interfere with the deposition process.

The disappearance of band 2a coincides with the first appearance of a band(s) corresponding to higher molecular weight μ-hydroxo bridged oligomers. These band(s) are firmly held on the chromatographic column requiring in some cases oxalate or even hydrochloric acid to elute them. The UV-visible λ_{max} increases in wavelength with continuing electrolysis, consistent with increasing molecular weight and number of chromium units per oligromer. It proved impractical to determine the concentration of these species by ICP spectroscopy, as these species elute in a very poorly resolved series of tailing bands. Their total relative concentration is significant but small (<10%).

Adsorption of these larger μ-hydroxo bridged oligomers onto the crystal growth site of the metal is believed to cause the loss of deposition rate and metallic quality. Methanol does not appear to adequately encourage their desorption but rather delay their first appearance. As adsorbants they could be very effective at relatively low concentrations.

In the presence of methanol and formic acid at pH 0.5–0.6 the first appearance of the higher oligomers and disappearance of green-blue band 2g is delayed until after 160 hours of electrolysis. In the presence of formic acid without methanol the green-blue band 2g has a much shorter lifetime and the appearance of the oligomers occurs sooner. Thus although much of the accelerated deposition rate can be achieved with addition of formic acid alone, the working lifetime of rapid deposition is much reduced. The addition of methanol alone is also less effective in prolonging rapid deposition than addition of both together. In the absence of formic acid and methanol the green-blue band 2g either does not appear or has an extremely short lifetime and higher oligomers are detected from the outset, consistent with the low rate of deposition of these original electrolytes.

Clearly methanol and formic acid are acting to stabilise desirable forms and hindering oligomerisation. Neither are acting as primary complexants. No bands were found corresponding to formate or methanol complexes. All the bands can be found in the absence of formic acid and methanol. (Even band 2g can be found in the absence of formic acid and methanol if the electrolyte is allowed to sit unused for a year).

The most likely role for the formic acid is as a complexant for the intermediate transient chromium(II). With formic acid the chromium(II) forms a quadruply bonded chromium(II)

Table 6 *The λ_{max} (nm) in the UV-visible spectra of the bands eluted from the chromatographic column, for the electrolyte with methanol and formic acid operating at pH 0.5*

Hrs Electrol.	0	1	24	48	96	120	144	168	192	216
Band 0	634									
	445									
Band 1	616	618	606	606	610	608	605	606	604	608
	433	434	428	427	428	427	426	426	426	427
Band 2		598	591	590	587	583	580	577	577	575
		422	419	417	416	413	413	408	409	408
Band 3			576	575	579	577	580	582	580	579
			417	416	418	417	417	417	417	414
Oxal. band							576	578	579	584
							419	419	418	422

Note bands 2b and 2g are not clearly resolved, the figures above represent the averaged λ_{max} of the mixture from the double overlapping band. At and after 168 hours only 2b is present.

Table 7 *The relative quantity of chromium, determined by ICP-AES spectroscopy, in each band as a fraction of the total chromium added to the chromatographic column, for the electrolyte with methanol and formic acid operating at pH 0.5*

Hrs Elec.	0	1	24	48	72	96	144	168	192	216	240
Band 0	0.78										
Band 1	0.22	0.79	0.26	0.27	0.38	0.33	0.29	0.38	0.39	0.36	0.22
Band 2b		0.17	0.50	0.37	0.34	0.47	0.44	0.54	0.53	0.51	0.64
Band 2g			0.20	0.31	0.24	0.14	0.20				
Band 3		0.04	0.05	0.06	0.04	0.06	0.06	0.06	0.07	0.05	0.07

Note bands 2b and 2g are not clearly resolved, the figures above represent the top and bottom of the double overlapping band.

dimer. The simple monomeric form of chromium(II) catalyses chromium(III) ligand exchange by forming a dimer with the chromium(III) as reaction intermediate. The quadruply bonded formatochromium(II) dimer would thus be a much less effective catalyst. Thus the formic acid would act to stabilise the critical species of band 2g and to hinder the oligomerisation. Larger quantities of formic acid would begin to form complexes with the bulk chromium(III) replacing other ligands of importance, explaining the loss of quality and deposition rate in this electrolyte with too high a quantity of formic acid.

The most likely role of the methanol is through the salvation sphere and its ability to hinder aggregation. Aggregation of the chromium species will be the first step in oligomerisation. Methanol has a well established role in hindering aggregation of dye stuffs etc. A similar role is likely for the aggregation of chromium species. It is significant that if methanol is replaced

Table 8 *A comparison of the UV-visible λ_{max} values of the chromatographic bands and values from the literature for known species*

Band	Observed λ_{max}	Assigned structure	Literature values	
			λ_{max}	Ref
0	634,445	$[Cr(H_2O)_4Cl_2]^+$	635,450	17
1	616,433*	$[Cr(H_2O)_5Cl]^{2+}$	608,428	18
2b	577,408	$[Cr(H_2O)_6]^{3+}$	575,408	9,19,21
2g	>590,>41*	$[Cr(Urea)_x(H_2O)_{6-x}]^{3+}$	-	-
3	579,417	$Cr(\mu–OH)_2Cr$	580,418	9,19,20
4	576–584	$Cr_3(\mu–OH)_4$	580,423	9,20,21
	419–422			
		$Cr_4(\mu–OH)_6$	580,426	9
		$Cr_6(\mu–OH)_{10}$	585,426	9

* The spectral shift is consistent with urea replacing some of the water ligands band 2g could not be separated from band 2b, the values above for 2g are an average of λ_{max} for 2b and 2g.

by ethanol, t-butanol or acetone, a similar but smaller acceleration of the deposition rate is observed. The order of effectiveness of the these solvents in this respect is in order of their solvation power and their ability to deaggregate dye stuffs. The strongest salvation is in methanol, the most effective solvent for this purpose. The strong salvation is also likely to slow chromium(III) ligand exchange to an even greater extent, increasing the stability of species present.

The role of pH is also critical. If the pH of the electrolyte containing formic acid and methanol is artificially raised to the make up pH 1.8, the oligomeric species appear quickly and high metallic quality and deposition rate are lost, even though the band 2g remains. In the electrolyte allowed to maintain its own pH, the species of band 2g would appear to create the low pH values necessary for methanol and formic acid to delay olation oligomerisation, while itself being stabilised by those two reagents.

4 CONCLUSION AND SUMMARY

A urea and chromium(III) based electrolyte has been developed from which an excellent quality of deposit and a rapid deposition rate can be obtained. Results showed that the presence of formic acid and methanol enhance both the quality and the deposition rate. In the presence of these electrolyte components, and very conventional conditions such as 25°C, no pH adjustment, low current density and very gentle agitation, an excellent quality of the deposit was obtained with a rapid deposition rate of 50–100 μm/hr. A sustained deposition of good quality was obtained up to 200 μm or above.

Ion exchange chromatography has demonstrated that the failure of the high deposition rate is associated with the first appearance of μ-hydroxobridged oligomeric species (olation) which are believed to adsorb on and poison the metal crystal growth site. Methanol and formic acid have been shown in this electrolyte to be particularly effective in delaying oligomer formation. Methanol is believed to act through solvation to hinder aggregation of the chromium species, while formic acid is believed to reduce the chromium(II) catalysis leading to electrolyte instability and olation. A very stable electrolyte has thus been achieved. The process has been assisted by a very low working pH associated with a particular electrolyte species.

A series of solvents were investigated for their ability to enhance the rate of deposition. It was found that the solvent should be a polar solvent to allow dissolution of the salts, have a high dielectric constant and high cohesive energy density but have NO donor capacity or ability to enter the coordination sphere of the chromium (III). Methanol has been shown to be the most suitable solvent. This enhancement of deposition rate was only observed in the presence of chloride ions and does not occur in the presence of sulphate ions which can also form stable bridged species.

The identity of the species in the electrolyte is being further confirmed by such techniques as differential polarography and FT-Infrared spectroscopy. The mechanical properties, corrosion and wear resistance properties of the coatings with respect to deposition **and** electrolyte parameters and in comparison to conventional coatings from chromium(VI) electrolytes is under investigation. This will be the basis of future publications.

Acknowledgement

The authors thank EPSRC for funding this project. Dr. S. K. Ibrahim also thanks the Centre for Particle Characterisation and Analysis at University of Paisley for the ICP analysis.

References

1. R. Bunsen, *Ann. phys.*, 1954, **91**, 619.
2. A Watson, A. M. H. Anderson, M. R. El-Sherif, and C. U. Chisholm, Trans. IMF 5 1990, 68, 26.
3. A. Watson, C. U. Chisholm and M. R. El-Sherif, *Trans. IMF*, 1986, **64**,149.
4. T. Hayaski and H. Nishikawa, *J. Met. Finish. Soc., Jap.*, 1974, **25**, 660.
5. C. Barnes, J. B. B. Ward and J. R. House, *Trans. IMF*, 1977, **55**, 73.
6. M. El-Sherif, S. Ma and C. U. Chisholm, *Trans. IMF*, 1995, **73**, 19.
7. A. S-M. Kasaaian and J. Dash, U.K. patent 2,1–54, 247A, September 4, 1985.
8. J. P. Benaben, 12th World Congress on Surface Finishing, Paris, October 1988.
9. H. Sttinzi and W. Marty, *Inorg. Chem.*, 1983, **22**, 2145.
10. T. Yoshida, "Process for chromium electrodeposition" U.S. patent 2,704.273, March 15, 1955.
11. J. J. B. Ward and C. Barnes, U.K. patent 1,488,381, October 12, 1977.
12. A. Smith, PhD, Thesis, 1994.
13. L. Glanclos, *Plating and Surface Finishing*, 1979, **66**, 56.
14. A. M. H. Anderson, PhD Thesis, 1990.
15. L. Tomio, E. Reddi, G. Jori, P. L. Zorat, G. B. Pizzi and F. Calzavara, *Springer Ser. Opt. Sci.*, 1980, **22**, 76.
16. A. Brenner, 'Electrodeposition of Alloys', Vol. 1, Academic Press, New York, 1965.
17. Kina, Wood and Gates, *J. American Chemistry Society*, 1958, **80**, 5015.
18. Tatibe and Myers, *J. American Chemistry Society*, 1954, **76**, 2103.
19. Laswick and Plane, *J. American Chemistry Society*, 1959, **81**, 3564.
20. Ardon and Plane, *J. American Chemistry Society*, 1959, **81**, 3197.
21. Thompson and Connick, *Inorganic Chemistry*, 1981, **20**, 2279.

2.5.2
Electrodeposition of Multilayer Zn-Ni Coatings From a Single Electrolyte

M. R. Kalantary[1], G. D. Wilcox and D. R. Gabe

IPTME, LOUGHBOROUGH UNIVERSITY OF TECHNOLOGY, LEICS., LEI I 3TU, UK

[1]NOW AT AVX LIMITED. LONG ROAD, PAIGNTON, DEVON. TQ4 7ER, UK

1 INTRODUCTION

In the last 15 years research on the electrodeposition of multilayered alloys such as Co-Cu, FeNi-Cu/Cu, Co-Ni-Cu/Cu, systems for electronic application, especially in the production of magnetic sensors for data recording, have been investigated[1-2] as possible alternative methods of production to that of physical vapour deposition (PVD). The demand for alternative processes is on the basis that the cost associated with vacuum technology is very high in comparison to the electroplating technique. Also, depending on the type of multilayer coatings produced, the latter can facilitate enhanced surface engineering properties; for example, Eckler et al[3]. claim the Co-Mo system used in gas turbine engines can improve wear, erosion and corrosion.

The process that involves the electrodeposition of multilayered alloys from a single bath (electrolyte) requires the application of a periodically varied current or potential. The use of a 'pulse' rectifier technique allows the deposition of layered coatings and, depending upon the type of electrolyte used, the composition of each layer of the coating can be in the form of either an almost pure metal or an alloy.

The control of composition of the deposit depends on the electrolyte system used. For example:

1. In the Cu–Ni system it is possible to assign the composition of each layer by careful control of electroplating parameters such as potential or current. At low overpotential or current the more electropositive metal deposits for the desired length of time without the deposition of the more electronegative metal. The purity of the deposit can be controlled by the concentration of the metal ions in solution and by agitation. In the Cu-Ni system the concentration is typically 1:100. When depositing copper at low overpotential or current, vigorous agitation is used to aid mass transfer of copper ions to the surface of the electrode and when plating nickel the agitation is removed so minimizing the copper content of the nickel layer.

2. In the Zn-Ni single bath system, it is not possible to produce a pure layer deposit because of the anomalous co-deposition mechanism associated with electroplating this type of alloy[5-9].

However, it is possible to produce a multilayer alloy coating having various concentrations of zinc and nickel in each layer. The change in composition at each layer can be controlled by current/potential, agitation and solution concentration. The production of multilayered coatings can be achieved by current or potential modulation, solution agitation, or a combination of the two. The study here is concerned with the production of multilayered alloy electrodeposits produced from a single bath using a pulse/periodic current modulation technique at constant electrode agitation/rotation. The major studies of the multilayered alloy coatings have been carried out on systems such as the Cu-Ni system that allow deposition of more pure layer deposit compositions. This has been carried out with a view to understand the mechanism of multilayer alloy deposition[10-11]. In this study, zinc-nickel is considered as a model system and the object has been to enhance the corrosion protection of steel substrates.

In order to produce multilayer alloy coatings of known composition and structure, the effects of current density and electrode agitation had to be established for single layer coatings, to obtain an operating window for surface appearance, composition, structure and corrosion resistance. The result of this work has been published elsewhere and some multilayer alloy coatings have also been produced by direct current methods with single and dual solutions to prove the production of CMA electrodeposits[12].

In this study pulsed current modulation has been used to produce compositionally modulated zinc-nickel alloy electrodeposits from a sulphate based non-complexing electrolyte. CMA coatings have been produced, having different composition and thickness at each layer. The corrosion performance of these multilayer alloy coatings has been assessed by neutral salt spray corrosion tests and compared with those for a single layer deposit.

2 EXPERIMENTAL

Iron cylinders (99.5%) with an outer diameter of 25 mm were used as the substrate material, Degreasing was carried out in an alkaline cleaner and then samples were etched in 50% v/v S.G. 1.18 hydrochloric acid for one minute at 23°C. Subsequently, after cleaning, electrodeposition trials were performed galvanostatically on the cylinder electrodes at 23°C in an electrolyte having, 35 g/1 zinc as zinc sulphate, 35 g/1 nickel as nickel sulphate and 80 g/1 sodium sulphate to increase the conductivity of the electrolyte. The electrolyte was adjusted to pH=2 using dilute sulphuric acid. A cylindrical stainless steel (AISI 316) foil was used as the counter electrode (anode). Compositionally modulated alloy electrodeposits having different layer thicknesses were produced by a current modulation technique. The overall coating thickness of all samples was eight microns. The sample identification and electrodeposition conditions are shown in Table 1.

As a basis for comparison of corrosion performance, one set of samples was electrodeposited at 5A/dm² in an electrolyte containing 35 g/1 zinc as zinc sulphate, 80 g/1 sodium sulphate at pH 2.5. A cylindrical zinc foil was used as the counter electrode (anode). Substrate (cathode) agitation was carried out by rotation at 100 r.p.m. The corrosion resistance of these samples was then compared to the single layer and multilayer (CMA) zinc-nickel coated samples mentioned earlier,

The compositionally modulated alloy coatings produced were also characterised by their top layer surface appearance, cross sectional and top surface composition and surface morphology. The surface morphology and compositional analysis of the CMA and single layer samples were carried out using a scanning electron microscope (SEM) and an associated

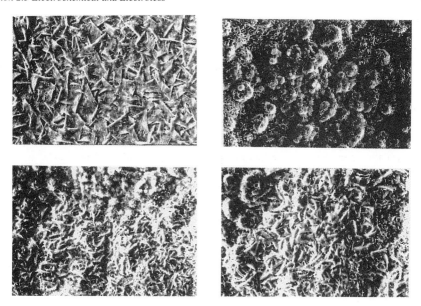

Figure 1 *Typical surface morphology of selected top surface of compositionally modulated alloy (CMA) coatings show that the deposit's top surface has similar characteristics to a single layer coating – clockwise from top left:10R8, 50R8, 5010R1, 1050R1*

energy dispersive X-ray microanalyser respectively. The corrosion tests were carried out in a 5 % wt/wt neutral salt spray according to ASTM B117.

3 RESULTS AND DISCUSSION

Electrodeposition of alloys by current modulation have been used for the production of the

Table 1 *Single layer and CMA coated sample identification*

10 R8	1 layer, microns in thickness
10/50 R4	2 layers each 4 microns in thickness
10/50 R2	4 layers each 2 miccrons in thickness
10/50 R1	8 layers each 1 micron in thickness
50 R8	1 layer, 8 microns in thickness
50/10 R4	2 layers each 4 microns in thickness
50/10 R2	4 layers each 2 microns in thickness
50/10 R1	8 layers each 1 micron in thickness

Note:
$10 = 10 A/dm^2$, $50 = 50 A/dm^2$ Electrode speed = 100 rpm
R4 = Each layer 4 microns in thickness
10/50 = Alternate layers were deposited starting at 10 A/dm^2
50/10 = Alternate layers were produced starting at 50 A/dm^2

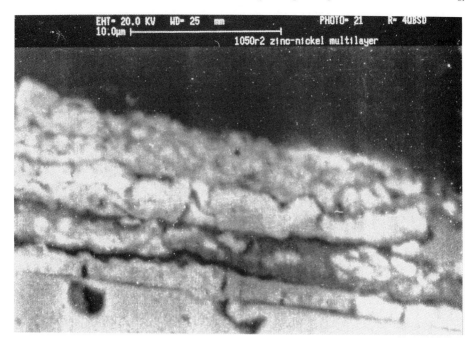

Figure 2 *Typical morphology of the cross-section of CMA coating showing four distinct layer (1050R2)*

compositionally modulated alloy (CMA) structures, the periodic modulation of current for the desired length of time achieving the appropriate thicknesses. It was thus possible to produce a multilayer alloy coating, having a different composition and structure at each layer. As will be described below, this has a considerable effect on the deposit properties in terms of corrosion resistance.

Table 2 *Nickel content of each layer of selected CMA coated samples*

Sample identification	Normalised values of nickel content (Wt. %)							
	Cross sectional analysis concentration of nickel at each layer, starting from the first layer adjacent to the substrate							
	Layer number							
	1	2	3	4	5	6	7	8
10 R8	8							
10/50 R4	8	20						
10/50 R2	9	19	15	16				
10/50 R1	13	24	10	26	10	20	10	20
50R8	18							

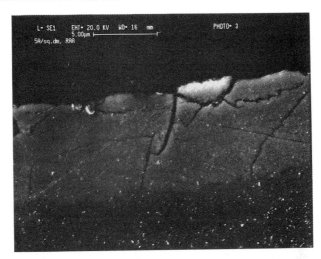

Figure 3 *Typical morphology of the cross-section of a single layer alloy coating (5RR)*

The surface appearance of the CMA coatings, having the same electrodeposition characteristics, has generally been found to be similar to that of a single layer coating (Figure 1). At low current density (10A/dm²) the deposit was found to be bright and at high current density (50A/dm²) the deposit appearance was found to have a dark grey colour. Present experience with CMA coatings, using the current modulation technique, proved that it is possible to control the deposit top surface appearance in terms of brightness or darkness by manipulation of current at the last stage of the electroplating procedure.

Table 3 *Corrosion test results of pure zinc deposit, simple alloy system and compositionally modulated alloys after the salt spray test in 5% wt/wt sodium chloride solution*

	Corrosion rate after salt spray test		
		ASTM B117	
Sample number	41 hrs	48 hrs	240 hrs
Zinc 5DC		50%RR	
10R8DC	WC		20%RR
1050R4	WC		5–10%RR
1050R2	WC		5–10%RR
1050R1	WC		< 5%RR
50R8DC	WC		20%RR
5010R4	WC		25%RR
5010R2	WC		10%RR
5010R1	WC		< 5%RR

RR – Red rust, WC – White corrosion, Zinc5DC – Zinc deposit alone, current density 5A/dm².

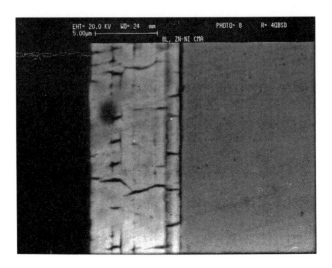

Figure 4 *Typical morphology of the cross-section of a CMA coating showing eight layers of electrodeposits (30/50)*

Previous work[12] on the simulation of high speed electrodeposition for the production of the zinc-nickel CMA coatings has revealed that there is little or no effect on deposit enhancement property that could be directly attributed to the surface morphology or texture. From the results obtained here (Figure 1) it is apparent that the composition of the alloy is the major factor influencing the deposit properties rather than the deposit morphology. The morphology of the cross-section of the zinc-nickel CMA coated sample in Figure 2 confirms that the deposit consists of layered structures. Further confirmation has been given below by the results of compositional analysis of the individual layers.

The morphology of the cross section of a zinc-nickel CMA coating shows (Figure 2) some cracks. This is characteristic of these zinc-nickel alloy coatings. The cross-sectional morphology of the samples produced from a proprietary solution for single layer and multilayered (CMA) coatings shows evidence of the cracks in the deposit (Figures 3 and 4). The number and the size of the cracks in the deposit can increase during preparation of the samples, since the deposit is very hard and brittle so during cutting, cracks will initiate at defect points or where there are pre-existing cracks, and later propagate. The presence of cracks in the deposit is due to high stress accumulated in the deposit, probably caused by the presence of oxide/hydroxide in the deposit[13] and which may also increase its hardness. However, the enhanced properties obtained from zinc-nickel alloy coatings in comparison to zinc deposits alone is sufficient to warrant increasing usage for selected applications, typically automotive bodies. In today's market, it is also general practice to apply zinc-nickel (or an alternative such as Zn-Co) for corrosion protection of nuts and bolts and other fastener parts[8,14].

The composition of the CMA coatings has a major effect on the overall deposit properties such as mechanical or corrosion performance[12]. For this reason a selected sample prepared by conventional metallography had each layer compositionally analysed. The result of the EDX analysis of the coatings is shown in Table 2; the results of the cross-sectional morphology

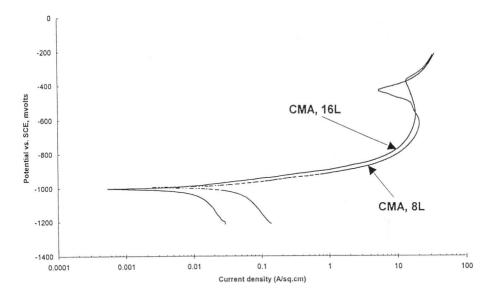

Figure 5 *Potentiostatic cathodic and anodic polarisation curves of CMA electrodeposits in 3.5% sodium chloride solution, sweep rate 30mV/min*

study of the sample confirms the production of a compositionally modulated alloy (CMA).

The results of the salt spray corrosion tests for the zinc only deposits, single layer zinc-nickel alloys and nickel CMA coated samples are shown in Table3. The corrosion test results for zinc-nickel alloy coatings were found to be superior to that of zinc alone. The single layer zinc alloy samples deposited at 10 A/dm² and 50 A/dm² produced 20% red rust after 240 hours whereas zinc alone produced 50% red rust after 48 hours of salt spray testing. 'The corrosion resistance of the single layer zinc–nickel alloy coatings produced at 10 A/dm² and 50 A/dm² showed similar corrosion performance. This agrees with the results reported elsewhere[8,12] which also indicate a preferred deposits composition of about 14% nickel.

From the results of the salt spray corrosion tests it is apparent that, in general, the CMA coatings produced a better corrosion resistance property in comparison to single layer coatings However, amongst all the layered coatings examined, the corrosion resistance of the duplex deposit, which had the high nickel-content layer adjacent to the substrate, showed an inferior performance over the single layer deposit. This suggests that a) for duplex thick layered coatings, the first layer adjacent to the substrate should be less noble, b) when the deposit consists of repeated layers the corrosion potential of the overall deposit has a less adverse effect than it would with a duplex layer which has a similar layer composition. From these results it can be concluded that for a given thickness the corrosion properties of the CMA coatings are similar and probably superior to their single layer counterparts.

The salt spray test results (Table 3) appeared to show a better result for the CMA coatings with a high number of repeated layers. Hence under the conditions studied the greater the number of layers in a CMA coating, the better the deposit's corrosion resistance. Although there is little or no known published work on the corrosion property of these types of CMA coatings, the conclusions here indicate promise. For future work CMA coatings containing a greater number of layers will be produced and the corrosion performance of the coatings will be assessed to see if they produce enhanced properties.

To further investigate the corrosion characteristics of the CMA coated samples, eight layer and sixteen layer specimens were subjected to cathodic and anodic polarization in 3.5% (by weight) NaCl solution (see Figure 5). It can be shown that the value of I_{corr} the corrosion current density, increases with increase in the number of alloy layers present. This appears to be mainly due to the effect of the two different structures on the cathodic reaction rate.

4 CONCLUSIONS

1. The surface appearance of the CMA coatings has been found to be similar to that of single layer coating and therefore typical of the electrodeposition conditions *per se*.
2. The morphology of the cross-section of CMA coated samples and the results of the Energy Dispersive X-ray Microanalysis (EDX), have proved the layered nature of the CMA coatings.
3. The CMA coatings, in general, produced better corrosion resistance than single layer deposits.
4. It has been found that, for the system and electrodeposition conditions employed the higher the number of repeated multilayers, the better the corrosion resistance of the total coating system.

Acknowlegements

This work has been funded as SERC Project No. GR/H 98007 in the UK.

References

1. M. Dariel, L. H Bennet, D. S. Lashmore, P. Lubitz, M. Rubinstein, W. L. Lechter and M. Z. Harford, *J Appl.Phys.*,1987, **61**, 4062.
2. W. Schwarzacher, M. Alper, K. Attenborough, P. Evans, R. Hart, G. Nabiyouni and D. S. Lashmore, Electrochem 94, 11–14th September 1994, Edinburgh.
3. T. A. Eckler, B. A. Manty and P L. McDaniel, *Plat. Surf. Fin.*,1980, **67**, 60.
4. J. P. Celis, A. Haseeb and J. R. Roos, *Trans. IMF*, 1992, **70**, 123.
5. A. Brenner, Electrodeposition of Alloys, Vols 1 and 2, Academic Press, New York and London,1963.
6. K. R. Baldwin, M. J. Robinson and C. .J Smith, *Trans. IMF*, 1994, **72**, 79.
7. S. A. Watson, Nickel Develpoment Institute Pub. No. 13001, 1988.
8. M. R. Kalantary, *Plat. Surf. Fin.*, 1994, **81**, 80.
9. G D Wilcox and D R Gabe, *Corrosion Science*, 1993, **35**, 1251–1258.
10. A. R. Despic and V. D. Jovic, *J. Electrochem. Soc.* 1987, **134**, 3004, and 1989, **136**, 1651.
11. A. Haseeb, J. P. Cells and J. R. Roos, *Transaction of Metal Finishers Association of India*, 1992, **1**, 15.
12. M. R. Kalantary, G. D. Wllcox and D R Gabe, *Electrochimica Acta*, 1995, **40** 1609.

13. M. R. Kalantary, G. D. Wilcox and D. R. Gabe, 'Auger electron spectroscopy and XPS studies of zinc-nickel alloy electrodeposits', Unpublished work, SERC project No. GR/H 98007 1995.

14. D. Smart and S. J. Wake, Proc. Symp, on Zinc Alloy Electrodeposits, IMF Midland Branch, Aston Uni., Birmingham, UK, 1988.

2.5.3
High Nickel Content Zinc-Nickel Electrodeposits from Sulphate Baths

C.E. Lehmberg, D.B. Lewis and G.W. Marshall

MATERIALS RESEARCH INSTITUTE, SHEFFIELD HALLAM UNIVERSITY, SHEFFIELD, UK

1 INTRODUCTION

To date the main interest in zinc-nickel electrodeposits has been in their use as corrosion resistant coatings[1-3]. The maximum corrosion resistance in such alloys when they contain between 12 and 13 mass percent nickel[4] and have γ-structures[5]. Work has been reported on acidic baths based upon both the mixed sulphates of zinc and a suitable alloying element, including nickel[6]. However little detail is available concerning baths suitable for producing good quality deposits of zinc alloys containing upwards of 25 mass % nickel or the structure

Table 1 *Composition and working parameters of nickel-zinc electrolytes*

Component	Solution							
	1	2	3	4	5	6	7	8
	mol l⁻¹	mol l⁻¹	mol l⁻¹	mol l⁻¹	mol l⁻¹	mol l⁻¹	mol l⁻¹	mol l⁻¹
$NiSO_4*6H_2O$	0.57	0.57	0.57	0.57	0.57	0.57	1.11	1.2
$ZnSO_4*7H_2O$	0.7	0.7	0.7	0.7	0.7	0.7	0.16	0.07
H_3BO_3	0.32	0.32	0.32	0.64	0.64	0.64	0.32	0.32
$C_7H_8O_3S*H_2O$	-	0.01	0.01	-	0.01	0.01	0.01	0.01
$NaC_{12}H_{25}SO_4$	-	-	0.002	-	-	0.002	-	-

Working conditions	Solution							
	1	2	3	4	5	6	7	8
Anodes	Nickel	Nickel	Nickel	Nickel	Nickel	Nickel	Inert	Inert
pH	3 (*/**)	3 (**)	3 (**)	~2 (***)	~2 (***)	~2 (***)	3 (**)	3 (**)
Volume	500 ml	500 ml	500 ml	500 ml	500 ml	500 ml	1500 ml	1500 ml
Agitation	a); b)	a); b)	a)	a); b)	a); b)	a)	b)	b)

*pH-value adjusted with ammonia solution
**pH-value adjusted with sodium hydroxide solution
***no pH-value adjusted; solution had ~2 agitation:
  a) submersible magnetic stirrer: speed 5, reverse mode 5
  b) agitated with nitrogen gas; 1.5 l min⁻¹

of such deposits. The aims of this work were therefore to investigate the production of high nickel concentration zinc alloys and the structural characteristics of such deposits.

2 EXPERIMENTAL

2.1 Plating Baths

A number of acidic plating baths each having a total metal ion concentration of 1.27 mol l⁻¹ but containing different ratios of zinc and nickel sulphates were subjected to the Hull Cell tests. These tests were used to determine the most suitable bath for use in the production of dense coherent deposits having high nickel contents. The solutions investigated in this way are listed in Table 1 along with the operating parameters used during preliminary tests.

Further experiments were done to produce electrodeposits with high nickel content at individual current densities using an electrolyte containing high nickel and low zinc concentration and additions of p–toluene sulphonic acid (solution 8 Table 1).

2.2 Hull Cell Tests

Preliminary tests were done at 1 ampere and 50 ºC using a 250 ml Hull Cell fitted with a platinized titanium anode and having a provision for gas agitation. Areas of deposits produced upon the Hull Cell panels were marked out along the horizontal plane at mid-height on the coatings and the composition of the alloys in these areas determined using energy dispersive X-ray Analysis (EDX). The current densities corresponding to the centre of each analysed area of the deposits were calculated as previously described[7].

2.3 Plating at Specific Current Densities

Electrolyses at specific current densities, in the range 10 to 60 mA cm⁻², were carried out in solution 8 (Table1) at 50 °C using copper cathodes placed centrally between two platinized titanium anodes. The cathodes were prepared using an abrasive cleaner, electrolytically cleaned, and activated prior to plating. The coatings produced in these runs were approximately 30 μm thick and was used for further compositional and structural studies. The composition of these were obtained using glow discharge optical emission spectroscopy (GDOES), the structures of the electrodeposits were analysed by X-ray diffraction using Cu-K$_\alpha$ radiation.

3 RESULTS

3.1 Preliminary Results

The Hull Cell tests showed that electrodeposits obtained from solutions 1 to 6 had relatively high zinc contents and tended to be non-coherent in the absence of p-toluene acid. Thus the zinc content of deposits obtained from baths 1 to 6 had values of between 80 and 90 mass % for current densities in the range of 5 to 50 mA cm⁻² but contained only about twenty mass % nickel.

Hull Cell panels obtained at overall currents of one and three amperes using plating solution number 8 without agitation produced only poor quality coatings on all areas of the test panels.

Table 2 *Results from Hull Cell tests using Ni–Zn solution number 8 agitated with nitrogen and a cell current of 1 A*

Current density (mA cm^{-2})	Composition		Observation
	Mass % Ni	Mass % Zn	Deposit appearance
4	63.0	36.9	Shiny; brownish-grey
5	60.7	39.3	Shiny; brownish-grey
15	45.9	54.1	Shiny; brownish-grey
20	32.1	67.9	Slightly dull; silverish-grey
30	26.4	73.6	Dull; silverish-grey
40	25.9	74.1	Dull; silverish-grey
50	44.9	55.1	Dull; silverish-grey

However tests repeated using nitrogen agitation showed that good coherent deposits could be obtained over a current density range from 4 to 50 mA cm^{-2} having nickel contents from around 30 to 75 mass % (see Table 2). However a slight increase of the zinc-nickel ratio in the electrolyte resulted in a remarkable increase of zinc-nickel ratio in the deposit (see Table 3).

3.2 Electrolyses at Specific Current Densities

The results obtained from the study of deposits obtained from solution 8 using current densities in the range 10 to 60 mA cm^{-2} and nitrogen gas agitation are summarised in Table 4.

The results confirm those obtained in preliminary work in regarding the variation of alloy composition with changes in cathode current density. Thus it can be seen from Table 4 that the zinc content of deposits first increases as the current density increased from 10 to 20 mA cm^{-2} before decreasing steadily as the current density increased further from 20 to 60 mA cm^{-2}. The nickel content of deposits as a function of current density can be seen to be more complex than the zinc content.

The results of structural investigations using X-ray diffraction given in Table 4 clearly show two different types of structure depending upon the current density used to produce the deposits. A single complex bcc γ-phase structure was obtained using current densities from 10 to 30 mA cm^{-2}. Figure 1 shows a typical X-ray diffraction pattern obtained from such a deposit. In contrast a more complex structure containing both bcc γ- and fcc α-phases was

Table 3 *Results from Hull Cell tests using Ni-Zn solution number 7 agitated with nitrogen and a cell current of 1 A*

Current density [mA cm^{-2}]	Composition		Observation
	Mass–% Ni	Mass–% Zn	Deposit appearance
5	38.8	61.2	shiny; brownish-grey
10	29.5	70.5	dull; brownish-grey
20	19.4	80.6	dull; yellowish-grey
30	19.0	81.0	shiny; silverish-grey
40	18.9	81.1	shiny; silverish-grey
50	19.8	80.2	shiny; silverish-grey

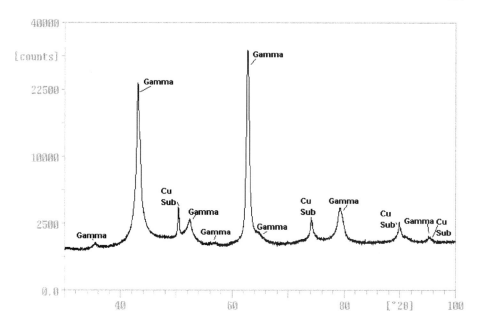

Figure 1 *X-ray diffraction trace of Ni-Zn electrodeposit with single phase: complex bcc γ-phase*

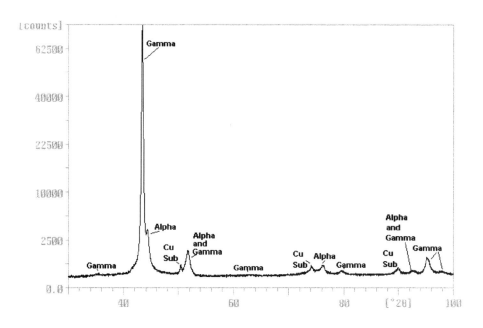

Figure 2 *X-ray diffraction trace of Ni-Zn electrodeposit with two phases: complex bcc γ- and fcc α-phase*

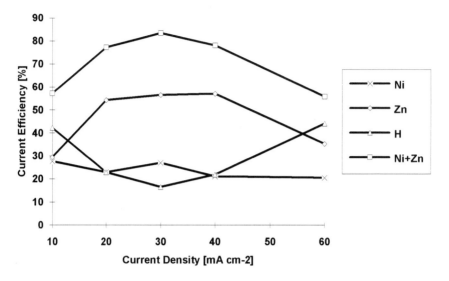

Figure 3 *Current efficiencies for nickel, zinc, hydrogen and alloy discharge as a function of current density*

obtained using current densities in the range 40 to 60 mA cm^{-2}. An X-ray diffraction trace obtained from this type of deposit is shown in Figure 2.

The overall and partial current efficiencies for nickel, zinc and hydrogen discharge calculated using the data from electrolyses carried out at specific current densities in conjunction with Faraday's laws are shown in Figure 3 as a function of cathode current density.

It can be seen that the overall current efficiency for alloy deposition peaks at 83 mass % at a current density of 30 mA cm^{-2}.

4 DISCUSSION AND CONCLUSIONS

Clearly dense coherent deposits of zinc-nickel alloys containing between 25 and 45 mass percent nickel can be obtained from an acidic sulphate bath containing 1.2 and 0.07 mol l^{-1} nickel and zinc respectively along with additions of boric and p-toluene sulphonic acids; and provided it is operated with good agitation and inert anodes.

The crystal structure of nickel is face centred cubic and that of zinc is hexagonal[8]. Without exception the nickel-zinc electrodeposits investigated showed the γ-phase, which is a complex body centred cubic lattice with 52 atoms per unit cell similar to γ-brass[9–11]. All investigated nickel-zinc electrodeposits showed the structure of the body centred cubic γ-phase, either as a single phase or as part of a mixed phase. Nickel–zinc electrodeposits prepared with higher current density but comparable nominal zinc compositions had the face centred cubic α-phase of nickel in addition to the body centred cubic γ-phase.

The occurrence of mixed face centred cubic α-phase and body centred cubic -phase appears not only to be a function of alloy composition, but also of the current density at which the alloys were electrodeposited. Thus in the nickel-zinc deposits obtained at higher current densities (i=40–60 mA cm^{-2}; Table 4), a mixed structure of face centred cubic α-phase and

Table 4 *Results of studies of Ni-Zn deposits obtained at current densities from 10 to 60 mA cm⁻²*

Current density (mA cm⁻²)	Composition Mass-% Ni	Mass-% Zn	Structure
10	45.8	54.2	bcc γ-phase
20	27.5	72.5	bcc γ-phase
30	30.0	70.0	bcc γ-phase
40	24.9	75.1	bcc γ-phase+fcc α-phase
50	28.5	71.5	bcc γ-phase+fcc α-phase
60	34.2	65.8	bcc γ-phase+fcc α-phase

body centred cubic γ-phase was identified. Deposits obtained from the same electrolyte at lower current densities (i=1–30 mA cm⁻²; Table 4) had similar or even higher nickel contents but only the body centred cubic γ-phase as a single phase.

The crystal structure of the nickel-zinc electrodeposits appears to be not only determined by the alloy composition, but also by the rate of electrocrystallisation. The higher the current density, the faster the electrodeposit is formed and the more equilibrium conditions are disturbed.

The relative efficiency for discharge of nickel, zinc and hydrogen during operation of the electrolytes varies as a function of current density. This efficiency of the zinc deposition increases with increasing current density up to the limiting current density. The highest efficiency of 57 mass % found for zinc deposition was at 40 mA cm⁻². Further increases in current density lead to lower zinc contents in the deposits since the cathode layer is depleted of zinc ions. The efficiency for nickel discharge generally shows a tendency to decrease as the current density increases. The overall current efficiency for alloy deposition increases as the current density increases until the limiting current density is reached at 30 mA cm⁻². The overall current efficiency for alloy deposition then gradually decreases as the current density increases further. The efficiency of the hydrogen discharge naturally behaves in the reverse manner. It decreases until the limiting current density for overall nickel-zinc alloy deposition is reached, and then increases as the current density increases further (see Figure 3).

The net result of these variations with operating current density is that careful monitoring and control would have to be exercised during any commercial operation of the bath.

References

1. L. Domnikow, *Metal Finishing*, 1965, **63**, 63.
2. A. Abibsi, J. K. Dennis and N. R. Short, *Trans. Inst. Metal. Finish.*, 1991, **69**, 145.
3. M. R. Kalantary, *Plating Surf. Finish.*, 1994, **81**, 80.
4. F. C. Walsh, *Trans. Inst. Metal Finish.*, 1991, **69**, 107.
5. British Patent 548,184, 1942.
6. S. R. Rajagopalan, *Metal Finishing*, 1972, **70**, 52.
7. DIN 50 957 'Galvanisierungsprüfung mit der Hullzelle', 1978.
8. Periodic System of Elements: Sargent-Welch Scientific Company, 1979.
9. W. Elkman, *Z. physik. Chem.*, 1931, **B12**, 69.
10. A. Westgren, *Z. Metallkunde*, 1930, **22**, 372.
11. J. Schramm, *Z. Metallkunde*, 1938, **30**, 122.

2.5.4
The Electrodeposition of Ni-P-SiC Deposits

D.B. Lewis, B.E. Dodds and G.W. Marshall

MATERIALS RESEARCH INSTITUTE, SHEFFIELD HALLAM UNIVERSITY, SHEFFIELD, UK

1 INTRODUCTION

The electroless deposition nickel-phosphorus coatings in both their basic form and as composites with ceramic dispersions is well established[1]. In addition composite coatings based upon single metal electrodeposits have been extensively studied[2-6]. Also Brenner et al. reported the electrodeposition of nickel phosphorus coatings[7] and although the process has had little impact on industrial practice a number of studies have been reported concerning the structure and properties of deposits obtained using the process[8-11].

Composite coatings based on electrodeposition provide an alternative coating to that of flame spraying for a wide range of engineering applications where corrosion and wear resistance are required[12].

Previous work has shown that nickel-phosphorus-alumina coatings can be produced by electrodeposition using sulphate/phosphate[13,14] and sulphamate[15] baths. The alumina however, was found to have a tendency to agglomerate and have a non-uniform distribution. This work reports further developments on the electrodeposition of Ni-P-SiC composites from a sulphamate bath.

2 EXPERIMENTAL PROCEDURE

2.1 Electrodeposition of Coatings

Prepared copper sheet specimens were positioned centrally between two anodes in the bath which was agitated vigorously using a combination of a reciprocating magnetic and two mechanical propeller stirrers see Figure 1 . The composition of the plating bath which was operated using cathode current densities between 10 and 80 mA cm^{-2} with and without the addition of 15 g l^{-1} silicon carbide was as follows:

NH_2SO_3H	49.5 g l^{-1}
H_3PO_3	0–45 g l^{-1}
$NiCO_3\,2Ni(OH)_2\,4H_2O$	42 g l^{-1}
$NiCl_2\,6H_2O$	45 g l^{-1}
$NiSO_4\,6H_2O$	148 g l^{-1}
Temperature	70 °C

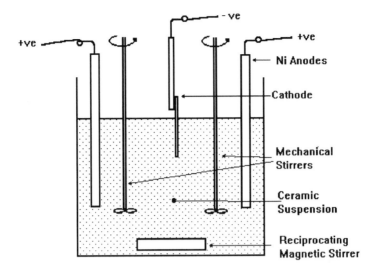

Figure 1 *Schematic diagram of bath for ceramic composite deposition*

The quantity of electricity passed during the course of each run was 1080 coulombs.

2.2 Heat Treatment

All heat treatment of coatings where appropriate was carried out for one hour duration at 400°C.

2.3 Examination of Coatings

The coatings were analysed using glow discharge optical emission spectroscopy (GDOES). Structural studies by X-ray diffraction using monochromatic C_u-K_α radiation were carried out in both the 'as-plated' and heat treated conditions.

Specimens containing silicon carbide were mounted in conducting bakelite and prepared in the usual manner to a one micron finish. The specimens were examined using scanning electron microscopy to assess the distribution of silicon carbide within the coatings using secondary and back scattered electron imaging techniques.

Microhardness measurements were made using a 25 g load and the final values obtained were based upon the average of eight indentations.

3 RESULTS

3.1 Cathode Current Efficiencies

Current efficiencies for the electrodeposition process were calculated using Faraday's Laws and the masses of nickel and phosphorus deposited . The valence of nickel and phosphorus were taken as two and three respectively for the purpose of these calculations. It was assumed however, for these calculations that the mechanism of silicon carbide inclusion involved a mechanical rather than a direct electrolytic process and therefore used no current directly.

A series of deposits from a bath containing 2.5 g l⁻¹ phosphorous acid with and without additions of silicon carbide and using current densities from 10 to 60 mA cm⁻² showed that the efficiency was relatively high over the whole current density range, between 90 and 97 per cent efficiency.

Figure 2 shows the partial efficiencies for nickel and phosphorus discharge, together with the overall efficiency for the bath as function of its phosphorous acid content when operated at 10 mA cm⁻². Superimposed on these results is the overall efficiency of the bath as a function of phosphorous acid content in the presence of a 15 g l⁻¹ suspension of silicon carbide.

3.2 Phosphorus Content of Deposits and Phosphorous Acid Content of Bath

The results of two series of plating runs carried out using cathode current densities of 10 mA cm⁻² in baths containing 15 g l⁻¹ and no silicon carbide respectively showed how the phosphorus content of the deposits obtained varied with the bath's phosphorous acid content.

The results of these experiments are shown in Figure 3 from which it can be seen that the phosphorus content of deposits obtained from the basic bath increased with the increasing acid content of baths. The relationship followed a similar trend for baths operated with and without silicon carbide suspensions.

3.3 Relationship between Operating Variables and Heat Treatment to Phosphorus Content of the Deposits

Two series of plated specimens were prepared to determine the effects of varying the phosphorous acid contents and the cathode current density used upon the silicon carbide and/ or phosphorus contents of deposits and hence their microhardnesses both before and following

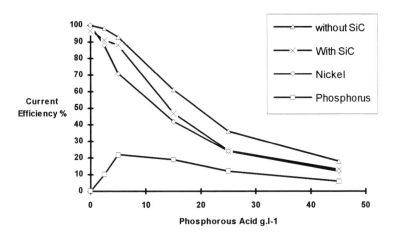

Figure 2 *The current efficiency of nickel, phosphorus as a function of phosphorous acid content and overall efficiency of the bath operated with and without a silicon carbide suspension*

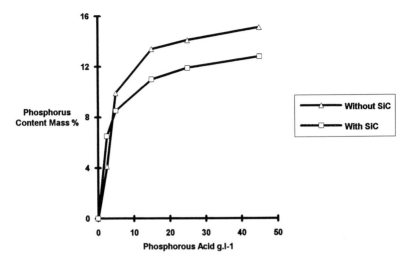

Figure 3 *Variation of phosphorus in deposits with change in phosphorous acid concentration*

heat treatment. The results obtained using the basic plating bath are given in Table 1 and those obtained using the bath operating with a suspension of 15 g l⁻¹ of silicon carbide are given in Table 2.

3.4 Metallographic Examination

Metallographic examination of coatings containing silicon carbide showed that the distribution of silicon carbide within the coatings was uniform as can be seen in Figure 4. This was in contrast to previous work on alumina which tended to agglomerate and have a non-uniform distribution[13–15].

Table 1 *Relationship between plating variables, the phosphorus content of deposits and their microhardness in the as-plated and heat treated condition*

H_3PO_3 gl⁻¹	Current density mA cm⁻²	Mass % P	Microhardness Hv25	
			As plated	Heat treated
0	10	0.0	250	247
2.5	10	4.13	382	871
5.0	10	9.9	436	862
15.0	10	13.4	366	720
25.0	10	14.1	416	706
45.0	10	15.1	447	650
2.5	20	3.0	364	720
2.5	40	2.8	435	495
2.5	60	1.53	421	475

Table 2 *Relationship between plating variables, the phosphorus content of deposits and their microhardness in the as-plated and heat treated condition for baths operated with silicon carbide suspensions*

H_3PO_3 gl^{-1}	Current density mA cm^{-2}	Mass % P	Mass %SiC	Microhardness Hv25	
				As plated	**Heat treated**
0	10	0	5	532	460
2.5	10	6.6	4.4	566	543
5.0	10	8.5	2.6	598	650
15.0	10	11.0	3.4	622	669
25.0	10	12.0	4.7	608	676
45.0	10	12.8	2.6	Insufficient thickness	Insufficient thickness
2.5	20	6.8	4.0	693	609
2.5	40	3.04	2.4	594	630

3.5 X-ray Diffraction

X-ray diffraction from coatings, of similar nominal composition deposited at 10 mA cm^{-2} from solutions with and without dispersions of silicon carbide are shown in Figures 5 and 6 respectively. It can be seen that both deposits had broad diffuse peaks which are typical of an amorphous structure. Coatings deposited from baths containing 15 gl^{-1} silicon carbide dispersions also showed sharp peaks corresponding to silicon carbide.

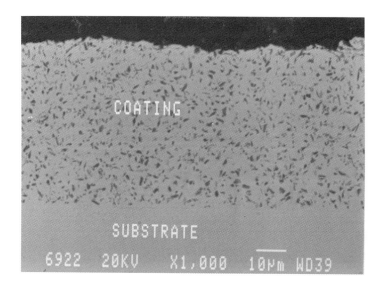

Figure 4 *Photomicrograph showing a Ni-P-SiC coating*

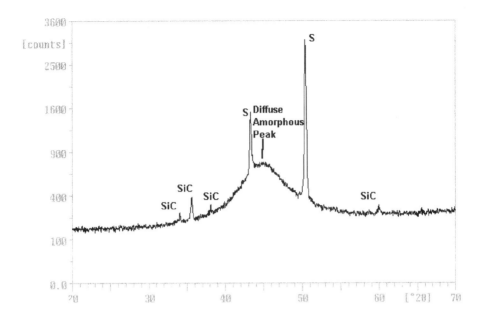

Figure 5 *X-ray diffraction trace of an as-deposited coating from a bath containing silicon carbide*

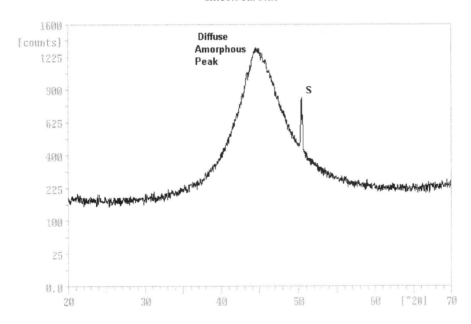

Figure 6 *X-ray diffraction trace of coating obtained from the basic solution containing no silicon carbide*

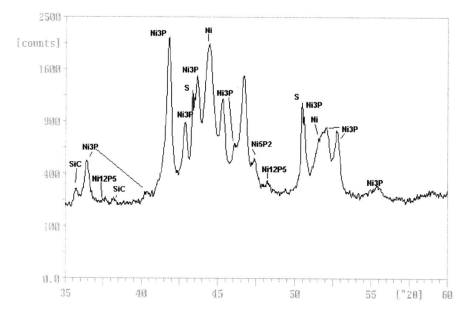

Figure 7 *X-ray diffraction trace of heat treated Ni-P-SiC Coating*

The X-ray diffraction traces obtained from coatings following heat treatment showed sharp well-defined peaks corresponding to crystalline structures. These structures containing mixtures of Ni_3P and Ni with and without silicon carbide can be seen from Figures 7 and 8 respectively. The deposits containing silicon carbide also contained the phases Ni_5P_2 and $Ni_{12}P_5$. The peak widths of Ni_3P and nickel, in terms of full width half maximum (FWHM), diffraction peaks in the deposits containing silicon carbide were also significantly broader than those without silicon carbide.

4 DISCUSSION

Figure 2 shows that not only does the overall current efficiency of the basic bath decrease with its increasing phosphorous acid content but that the variation in the efficiencies for phosphorus and nickel discharge do not follow the same trends. The consequence of this is that the ratio of phosphorus to nickel found in deposits obtained from bath containing increasing amounts of acid shows a maximum value. The overall change in efficiency with increasing phosphorous acid content of the bath shows a similar trend when the bath is operated with or without a silicon carbide suspension. However, one important difference is that generally the efficiencies obtained with silicon carbide in the bath are generally lower than those for the basic solution. The explanation for this may be simply that the efficiencies quoted are based upon the mass of the adherent deposits obtained. These may be lower in a bath operated with a ceramic suspension simply due to abrasive wear of the growing deposit by the silicon carbide simultaneous to the deposition process[16].

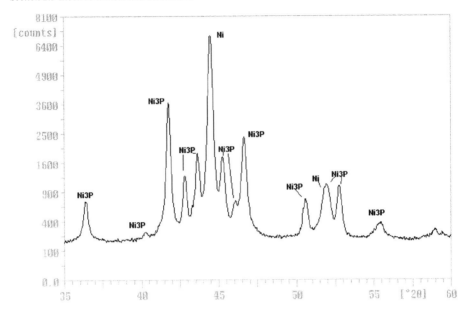

Figure 8 *X-ray diffraction trace of heat treated Ni-P coating*

Figure 2 shows that there is an increase in the phosphorus content of deposits obtained with increases in the phosphorous acid for a bath operated both with or without silicon carbide. It is worth noting that this relationship means that the phosphorus content of deposits obtained by electrodeposition could be controlled independently by varying the bath's phosphorous acid content.

Consideration of the data in Tables 1 and 2 suggest that the hardness of 'as-plated' and heat treated deposits obtained from baths operated both with and without silicon carbide has both important similarities and differences.

The 'as-plated' deposits containing both silicon carbide and phosphorus are harder than those obtained from the corresponding bath operated without ceramic suspensions (see Tables 1 and 2). This is probably due to a dispersion hardening effect of the harder silicon carbide particles in the 'as-plated' alloy matrix[17]. The peak in maximum hardness as a function of phosphorus content obtained following heat treatment of 'as-deposited' coatings from the basic plating solution also reflects the general behaviour of nickel phosphorus deposits previously noted[1]. In the heat treated condition the hardness of the deposits without silicon carbide were always harder than those containing silicon carbide. Thus, it can be seen, that, whereas deposits obtained from baths containing 5 to 25 gl^{-1} phosphorous acid without silicon carbide had hardnesses of 706 to 862 following heat treatment (see Table 1), those obtained from similar baths containing silicon carbide, which produced deposits having similar phosphorus contents but with additional silicon carbide, had hardnesses of about 650 to 676 following heat treatment (see Table 2). The practical implication of this might be that deposits containing silicon carbide may have better wear resistance than those without silicon carbide in the 'as-plated' condition but that the situation might be reversed in the case of the heat treated coatings.

The X-ray diffraction results confirm that the hardening observed when nickel-phosphorus electrodeposited coatings from the basic bath are heat treated results from a crystallisation process during which mixtures of Ni_3P and Ni are formed. This mechanism has previously been established in relation to such deposits obtained by electrodeposition[13], electroless deposition[1] and electroless deposition with superimposed pulsed current[11]. However, examination of the X-ray traces from coatings with and without silicon carbide dispersions show important differences. The deposits with SiC dispersions contained the metastable phases Ni_5P_2 and $Ni_{12}P_5$ in addition to the stable Ni and Ni_3P phases. The peak widths of the Ni_3P and Ni phases were significantly broader in the deposits containing silicon carbide dispersions. Previous findings[11] have shown that peak broadening and the presence of metastable phase was associated with the early stages of crystallisation, which occurs at a stage well before peak hardness develops. Thus, the lack of hardening response after heat treatment in the coatings containing silicon carbide dispersions is because the deposit is at an earlier stage of crystallisation.

5 CONCLUSIONS

1. Coherent nickel-phosphorus and nickel-phosphorus-silicon carbide coatings can be produced by electrodeposition from a bath containing phosphorous acid.
2. Unlike in the case of electroless deposition, the phosphorus content of nickel-phosphorus deposits obtained by electrodeposition can be controlled independently by simply varying the phosphorous acid content of the bath.
3. In the 'as-plated' condition electrodeposited coatings containing silicon carbide are harder than the corresponding coatings containing no silicon carbide.
4. Following heat treatment nickel-phosphorus coatings are harder than the corresponding deposits containing silicon carbide.
5. The presence of a silicon carbide suspension in baths containing more than 15 g l^{-1} phosphorous acid considerably reduces the ratio of phosphorus to nickel in the deposits obtained using a cathode current density of 10 mA cm^{-2}.

References

1.   W. Riedel, *Funktionelle Chemische Vernicklung*, Eugen G Leuze Verlag, Saulgau/ Germany, 1989.
2.   W. Metzger and Th. Florian, *Metalloberfläche*, 1980, **24**, 274.
3.   F. Sauter, *Metall*, 1964, **19**, 596.
4.   V. P. Greco and W. Baldauf, *Plating*, 1968, **3**, 250.
5.   E. Broszeit et al., *Metall*, 1971, **25**, 470.
6.   J. .P Celis et al., *Trans. Inst. Metal Finish.*, 1991, **69**, 133.
7.   A. Brenner et al., Research Paper RP2061 NBS, 1950, **44**, 109.
8.   A. Brenner, *Electrodeposition of Alloys*, Acadmic Press, New York, 1963.
9.   U. Pittman and S Ripper, *Zeitschrift Metallkunde*, 1983, **74**, 783.
10. D. S. Lashmore and J Weinroth, Second International Pulse Plating Symposium, Rosemont/USA, 6–7 October 1981.
11. G. W. Marshall et al., *Surface and Coating Technology*, 1992, **53**, 223.

12. A. R. Poeton, *Metals and Materials,* 1988, **4**, 702.
13. Editors P. K. Datta and J. S. Gray, 'Surface Engineering', Royal Society of Chemistry, Cambridge, 1993, Vol. 1, p. 225.
14. The Electrodeposition of Ni-P-Al$_2$O$_3$ Deposits, G.W Marshall, D. B. Lewis, D. Clayton, K. Blake and B. E. Dodds, in 'Proc. Int. Conf. on Advances in Materials and Processing Technologies', Vol III pp 1502–1511, 1995. Ed. M.S. J. Hashmi, Pub. Dublin City University.
15. H. Jacobs, D. B. Lewis and G. W. Marshall, Unpublished work.
16. M. R. Kalantary, K. A. Holbrook and P. B. Wells, *Trans. Inst. Metal Finish.,* 1993, **71**, 55.
17. J. C. Fisher, E. W. Hart and R. H. Pry, *Acta Met.,* 1953, **1**, 336.

2.5.5
Electroplated Iron Layers Containing Nitrogen

V. A. Stoyanov[1], J. D. Di-agieva[2] and A.I.Pavlikianova[1]

[1]THE UNIVERSITY OF ROUSSE, BULGARIA

[2]BULGARIAN ACADEMY OF SCIENCE, BULGARIA

1 INTRODUCTION

The initial applications of electroplated iron have been in the fields of electroforming and rebuilding of worn-out and incorrectly machined parts[1-3]. The interest in using electroplated irons in engineering is still very strong.

The reasons are several:

1. The electroplated iron is cheap and by its physico-chemical properties it approaches the middle-carbon steels. By thermal treatment its properties can be changed[3,4]. Its surface can be carbonized, nitrided and this improves its quality, including its contact fatigue strength[5].
2. The electrochemical process is a highly efficient method. The process is characterised by high current densities in the range of $10–100$ A/dm^2 and a high plating rate of up to 0.5 mm/h, that exceeds $15–20$ times the rate of chromium plating.
3. The electroplated iron layers differ from the metallurgical iron by properties and structure. They have a small-grain structure with a deformed crystal lattice. An important advantage used nowadays in the automobile industry stems from the ability of the electrolytic iron to absorb a strong oil film, facilitating its working under severe conditions, such as gears, cams, pistons etc[6-8].

To be used widely, the electrolytic iron has to possess the high requirements needed for the application in machine engineering layers. The mechanical properties hardness, wear protectiveness and internal stresses are some of the most important characteristics. They are influenced by various process variables, e.g. current density or cathode potential, eleclrolyte temperature, concentration of metal ions, other cations, anions, pH, inhibitors, complexing ligands.

The efforts of different electrochemical research laboratories are directed towards obtaining of electroplated iron layers, characterised not only by their purity but also by their microcrystal structures. This includes epitaxial layers, powders and various non-equilibrium structures, such as metastable phases, compositionally graded and modulated microstructures[9-11].

The aim of this investigation is to show the possibility to include nitrogen in the iron layer's crystal lattice during the process of electroplating. The desire is to obtain nitrided electroplated

(a)   (b)

Figure 1 *Heavy iron deposits from (a) $FeCl_2\ 4H_2O$ and (b) with TEA additive*

iron layer without additional thermochemical treatments as it has been done 'till now. The advantages can be several. One of them is that there will be no limits for the nitriding layer. Another expected result is an improved wear-protectiveness.

The idea is to carry out both the electrochemical plating process and chemical process, allowing the addition of nitrogen in the crystal lattice. Similar processes are carried out by plating of nickel phosphorous layers where sodium hypophosphate is used as a reducing agent[12].

Similarly, by the addition of small quantities of Triethanolamine (TEA) to the electrolytes for iron plating one can obtain layers containing nitrogen in the crystal lattice.

Table 1 *Influence of the electrolysis conditions on the N_2, O_2 and H_2 contents in the layer*

Solution	300 g/l $FeCl_2.4H_2O$		
Sample	**1**	**2**	**3**
TEA-additive (g/l)	0.02	0.02	-
t°C	20	50	70
jk (A/dm^2)	10	10	10
Rate (m/h)	100	100	100
pH	0.7	0.7	0.7
H_2 (% mass)	0.116	0.0367	0.018
O_2 (% mass)	2.257	0.236	0.228
N_2 (% mass)			
Gas analysis	0.0301	0.0096	0.0021
Keldahl	0.028	0.007	-

2 EXPERIMENTAL PROCEDURE

The electroplating of iron layers was carried out in chloride baths containing 300 g/l $FeCl_2.4H_2O$. The process of electrolysis was carried out at temperatures 20–70°C and containing three valent iron ions (Fe^{3+}) up to 1g/l. The investigations were carried out at a cathode current density (jk) in the range 10–30 A/dm² and pH in the ranges 0.5–2.5. Low carbon steel with an area of 100 mm² was used as a substrate. Plating was carried out in an electolyte with a volume of 5 l. For the introduction of nitrogen tests were carried out in a working bath with a capacity of 220 l.

The preliminary substrate preparation consisted of mechanical cleaning, chemical degreasing, washing , anodising for 30 s in 30 % H_2SO_4 at anode current density (ja) 30 A/dm².

The TEA was introduced into the electrolyte as an additive as a salt (hydrochloride) in quantities 0.02–0.4 g/l.

The layers' morphology, microhardness and microhomogeneity were determined. Microhardness is automatically determined by Vickers with a 100 g load for 5 s on a Shimadzu Hv tester.

The metal structure was investigated by an X-Ray diffractomeric and Mössbauer analysis. The quantity of the non-metal additives in the layers, nitrogen, oxygen and hydrogen, were also determined.

The layer samples were subjected to a gas analysis for nitrogen and oxygen by melting in vacuum by heating up to 1600°C (Exholograph EA-1, Balzers). Additionally, nitrogen was chemically determined in a carrier gas flow using LECORN device.

Only the temperature curves for nitrogen evolution were encountered on LECO-TN-314.

3 RESULTS AND DISCUSSIONS

It was initially established that the addition of TEA to the chloride iron plating electrolyte in quantities bigger than 0.1 g/l leads to the obtaining of rough and very brittle layers, not applicable for plating of parts. Additives in the ranges of 0.01–0.05 g/l allowed to obtain unusually thick (up to 5 mm) and smooth layers, compared to that obtained from the pure electrolyte (Figure 1). Therefore during the subsequent investigations 0.02 gl/l additive of TEA was used for this

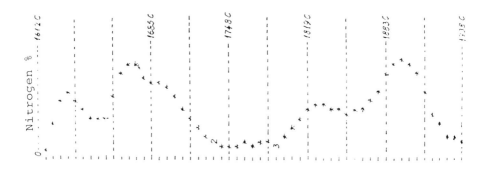

Figure 2 *Spectrogram giving temperature profile of the nitrogen given off from the iron coating obtained from the chloride electrolyte with TEA additive*

type of electrolyte. Selected results for this are summarised in Tables 1 and 2. Comparing the results from the gas analysis (Table 1) with the conditions for obtaining different samples, one concludes:

For the same metal-ion contents of the dissolved iron salt, the layers obtained with additives of TEA in the electrolyte (Samples 1 and 2) contain nitrogen. However, they also contain considerable quantities of hydrogen and oxygen. Increasing the temperature without altering other conditions, the inclusion of hydrogen and oxygen in the layer sharply decreases (samples 1 and 2 obtained by the presence of TEA in the electrolyte and Sample 3 – obtained by the absence of the additive). The quantity of nitrogen is highest in a layer obtained at 20°C (Sample 1). In the layer obtained by the presence of the same quantities of TEA but by increasing the temperature up to 50°C, the quantity of nitrogen becomes equal to the minimum value determined for conditions with no additive (Sample 3). It is obvious that the favourable conditions for the presence of nitrogen in the layers are the low working temperatures.

The explanation of the results can be found in the forming characteristics of the electroplated iron at the cathode. At low temperatures the coating becomes saturated more considerably with hydrogen and oxygen. The inclusion of hydroxides, appearing under these conditions of electrolysis at the by-cathode layer, has a strong influence as well. Hence, the state of nitrogen in the layer is of a special interest – it is either absorbed in the TEA composition or it is bonded to the iron crystal lattice.

Figure 2 shows the temperature profile of nitrogen given off (in mass %) from the iron layer by LECO-TN-314 for given intervals of time.

The maxima observed up to 1650°C are typical for absorbable included nitrogen while the observed maximum about 1894°C proves the presence of nitrogen bonded in the 1894°C metal iron lattice as a nitride. This leads to the necessity to think again over the TEA participation as a reducing agent giving electrons from the nitrogen atom of the amino group and introducing a nitrogen atom in the crystal iron lattice. Definitely, the absorption of TEA in the form of complex compounds with the three-valence iron on the cathode probably plays a considerable role. The high working temperatures lead to the desorption of TEA, as well as to the sharp decrease of its role as a reducing agent.

Another evidence supporting the inclusion of nitrogen in the crystal lattice of the electroplated iron is the increased microhardness and microhomogeneity of the layers obtained by the addition of TEA. Table 2 presents the results of the microhardness and microhomogeneity of layers obtained with or without addition of TEA. The dispersing of microhardness around 35% from point into point can be looked upon as microphase differences caused by the influence of TEA.

The X-Ray Diffractometric analaysis investigations of the layer with a hardness of 1000 HV showed the presence of a microcrystal and an X-Ray amorphous state of the layer. The dimensions of the crystals are considered to be about 140 Å (see Figure 3).

Table 2 *Microhardness and microhomogeneity of an electroplated iron layer*

Solution	No of measurements	Average microhardness	Microhomogeneity
$FeCl_2\,4H_2O$	9	400	25
$FeCl_2\,4H_2O$ (0.02 g/l TEA)	15	953.4	35

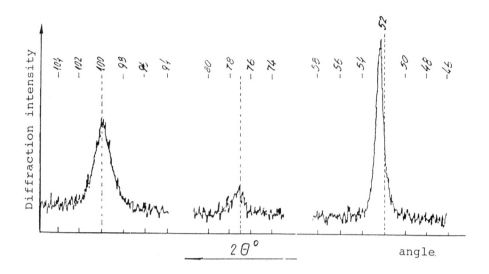

Figure 3 *Radiogram obtained by X-ray diffractometric investigation of the iron coating structure plated from chloride electrolyte with TEA additive*

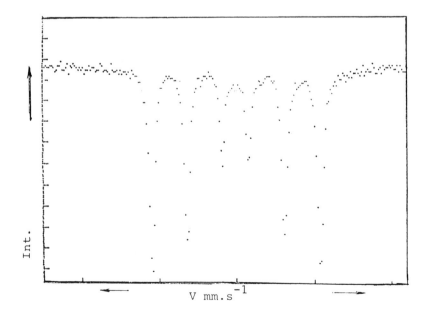

Figure 4 *Mössbauer spectrum of an iron coating with a 0.03 mass % of nitrogen*

The Mössbauer analysis of the layers containing TEA confirm the X-Ray amorphous scale. A phase has been observed in the Mössbauer spectrum of the iron (Figure 4), identified as containing nitrogen.

4 CONCLUSIONS

The results obtained demonstrated a way to carry out nitriding during the electroplating process of iron layers.

TEA, as an additive to the chloride electrolyte for electroplating of iron, is the nitrogen-containing compound to be used for the formation of thick nitrided layers with a microcrystal structure.

References

1. W. E. Hugas, Modern electroplating. A guide book for platers, workers, chemists and engineers, London, 1923.
2. C. T. Thornas and Blum, *Transactions of the American Electrochemical Society*, 1923, **59**, 59.
3. M. P. Melkov, A. N. Shvelzov, I. M. Melkova, 'Rebuilding of Automobile Parts by Hard Iron', Moscow, Transport, 1982.
4. R. Walker and S. D. Irvine, *Metal Finishing*, June, 1976, 39.
5. D. Scott and P. J. McCullagh, Hardenable electrodeposited cotings for rolling bearings – A preliminary assessment, *Electrodepos. Surface Treatment*, 1972/73, **1** , 21.
6. R. K. Kepple, E. R. Manlel, O. J. Klingenmaier, R.L. Mailson, *Journal of Lubrication Technology*, October, 1970, 557.
7. British Patent N 1232426, F16 and 1/00.
8. USA Patent N 3753664, Hard Iron Electroplated and Soft Substrates and Resultant Product.
9. 'New Trends and Approaches in Electrochemical Technology', Edited by Noborn Masuko, Tetsuya Osaka and Yashiro Fukunova, 1992.
10. USA Patent N 4388379, Electrodeposition of Low Stress Hard Iron Alloy and Article so Produced.
11. J. D. Thomas, O. J. Klingenmaier and D .W. Hardesly, *Transactions of the Institute of Metal Finishing*, 1969, **47**, 209.
12. K. M. Gorbunova, A. A. Nikiforova, Physico-Chemical foundations of the chemical nickel plating processes, Moscow Academny of Science, Russia, 1960.

2.5.6

The Effects of Bath Ageing on the Internal Stress Within Electroless Nickel Deposits and Other Factors Influenced by the Ageing Process

C. Kerr, D. Barker* and F. Walsh

SCHOOL OF CHEMISTRY, PHYSICS & RADIOGRAPHY, UNIVERSITY OF PORTSMOUTH, ST. MICHAEL'S BUILDING, WHITE SWAN ROAD, PORTSMOUTH, UK

*AUTHOR FOR CORRESPONDENCE.

1 INTRODUCTION

The ever increasing use of electroless nickel coatings over the last 20 years has been due to the unique combination of its properties such as corrosion resistance, hardness and wear resistance. All these properties make this coating ideal for a wide range of applications[1]. Problems, however, can arise if strict control over plating and bath conditions are not properly maintained which may result in pores or cracks developing in the coating. Examples of these include ; (i) pretreatment, (ii) substrate condition, (iii) surface roughness and (iv) residual or internal stresses. Any one of these, or combination of two or more, may contribute to the formation of defects in the coating. Of the factors mentioned, pretreatment and surface roughness[2-6] have been discussed in earlier papers and the aim of the present paper is to examine the effect of internal stress on the porosity of electroless nickel coatings.

2 LITERATURE REVIEW

Internal stress can either be tensile, compressive or zero. The nature of this internal or residual stress is of importance as the physical properties of the deposit may be undermined. For example, a coating with a high internal stress is hazardous to cathodic deposits as they can lead to crack formation and expose the substrate to the attacking environment, or peel leading to poor adhesion of the coating. Therefore, in early stages of deposit growth, the nature of the stress may have an important bearing on the final levels of coating porosity found within the plated deposit. Alternatively, a compressive stress[7] may not only help reduce porosity but also improve deposit adhesion as electroless nickel coatings tend to have a relatively low ductility, i.e., elongation of < 1 %. If there was a surface defect or soil residue on the surface of the substrate and the electroless nickel coating had high tensile internal stress, the deposit over the soil may peel off resulting in poor adhesion. However, if the coating possessed a compressive internal stress, the force of the deposit may push down at the site of the defect or soil, thereby, reducing the chance of coating failure as a result of poor adhesion.

In situations where the component is subjected to applied loads, the internal stresses within the deposit may have an important influence on the time to failure such as in stress corrosion cracking and corrosion fatigue conditions. The presence of a high tensile stress within the coating may be detrimental to both deposit and substrate alike[8]. The fatigue strength of a steel surface coated with an electroless nickel with a high internal tensile stress may be significantly reduced whereas a compressive stress may increase the fatigue strength of the metal. The tensile nature of the stress may make the metal of the coating more reactive (or anodic to the substrate) than a stress free deposit because of a higher internal energy. The stressed areas are, therefore, thermodynamically more likely to corrode.

The phosphorus content of the coating also affects the internal stress of electroless nickel deposits. A phosphorus content of between 11–13 % is said to result in a compressive stressed deposit. However, deposits with lower phosphorus content produce a tensile internal stress. Baldwin and Such reported[9] that the cause and effect of internal stress within electroless nickel deposits were a result of physical changes in the electroless bath, e.g., pH, temperature and phosphite concentration. On varying the pH, their results indicated a change from compressive to tensile internal stress at pH 4.65. They suggested that this was due to an increase in orthophosphite concentration as the bath aged. For example, an orthophosphite of 120 g dm^{-3} in the bath gave an internal stress of 117 MPa. These results underline the importance of controlling the bath ingredients and operating parameters thereby ensuring optimum plating conditions are maintained.

A freshly made up electroless nickel plating bath produces deposits with internal compressive stresses, the level of this stress decreases with phosphorus content until, eventually, the stress found within the plated deposit becomes tensile. The amount of phosphorus deposited in the coating is related to the age of the plating solution with levels of deposit phosphorus falling as the number of plated metal turnovers increases. A 1.0 metal turnover is equivalent to the plating out of 6.1 g dm^{-3} nickel metal from an electroless nickel plating solution if the initial make up contains a 6.1 g dm^{-3} nickel content.

Published internal stress measurements should only be used as a guide as there are numerous experimental methods employed to measure this property[10]. These include Brenner and Senderoff's spiral contractometer, Kushner's stressometer[11], flexible strip method and hole drilling. A further problem is type of substrate material used in the test as this can influence the level of stress recorded. For example, the stainless steel spirals used in the spiral contractometer method may give different results to a carbon steel one.

3 EXPERIMENTAL

10×2 cm samples were cut from 1.5 mm thick Pyrene steel plates. These were weighed and subjected to the following pretreatment prior to electroless nickel plating :-

1. Alkaline soak cleaner, (Duraprep 115, 25 g dm^{-3}). The panels were immersed for 5–10 minutes at a temperature of 55–60°C.
2. Rinse $\times$ 2, (deionised water). The panels were immersed for 30-60 seconds at a temperature of 22°C.
3. Pickling - activation, (hydrochloric acid S.G. 1.18, 25 % v/v). The panels were immersed for 60 seconds at a temperature of 22°C.
4. Rinse $\times$ 2.
5. Steps 1–4 repeated
6. Electroless nickel plating solution, (Duraposit 90). The panels were immersed at a temperature of 90–92°C and at a pH of 4.6–4.8. Different immersion times were employed and the variation of coating thickness with this parameter is shown in Table 1.

The overall plating rate for this electroless solution was approximately 14 mm hr^{-1} for a freshly prepared bath. The electroless nickel plated samples used in the electrochemical tests were fully covered with an acid resistant tape (Advanced tapes – AT163) exposing 1 cm^2 of the coated steel.

3.1 Plating Rate Determination

The Pyrene steel panels were pre-weighed prior to plating for 60 minutes in various electroless nickel solutions (the aged plating baths varied from 0-6 metal turnovers) and subsequently reweighed after plating.

3.2 Corrosion Potential (E$_{cor}$) Measurements

The sample electrode (various electroless nickel plated deposits) were immersed for 5 minutes in a 0.125M H$_2$SO$_4$ electrolyte at a temperature of 22°C to allow equilibrium conditions to become steady. The corrosion potential was measured against a saturated calomel reference electrode (S.C.E.) using a high impedance Fluke 8010A digital multimeter.

3.3 Polarisation Studies

For the anodic polarisation scans, a two compartment cell was used with the working electrode (WE) and a 3.4 cm^2 platinum foil counter electrode (CE) separated by a Nafion 324 membrane. The cation membrane separation prevented any products generated at the CE interfering with the electrochemical measurements. The S.C.E. reference electrode was connected to the cell via a Luggin capillary (the distance between the WE-Luggin capillary was approximately 1 mm). An EG & G potentiostat model 273 interfaced by a National Instruments via a GPIB-PCII board, running NI-488.2 software was used to produce current vs potential data on a 386 PC microcomputer. The taped sample was immersed into the electrolyte and initially scanned to a potential –250 mV more negative than its E$_{cor}$ potential. This ensured that the surface was cleaned prior to anodic polarisation. Once cleaned, the sample was scanned to a potential 3 V more positive than its E$_{cor}$ value. The scan rate adopted for the anodic polarisation scans and Tafel plots was 1 mV s^{-1}.

The cell arrangement and sample preparation to obtain Tafel data was similar to that used for the anodic polarisation scans except that the potential was scanned only to values 250 mV each side of the E$_{cor}$ value.

3.4 Internal or Residual Stress Measurements

The stress within an electroless nickel deposit was determined by the use of a Brenner-Senderoff spiral contractometer[12]. This instrument consisted of a strip of stainless steel (2.54 x 30 x 0.15 cm thick) wound in the shape of a helix (2.5 cm diameter). Its interior was coated with a lacquer thus preventing the deposition of electroless nickel inside of the helix. The helix was fixed at one end of the spiral contractometer while the other end was free to move, thereby, moving a pointer on the dial. From the direction of rotation (clockwise for tensile and anti-clockwise for compressive stress) and amount of deflection recorded, the type and extent of internal stress was then measured.

The stainless steel spirals were chemically treated prior to electroless nickel plating as follows (with times and conditions as for the mild steel pretreatment):

1. Alkaline soak cleaner, (Duraprep 115).
2. Rinse × 2, (deionised water).
3. Pickling, (hydrochloric acid).
4. Rinse × 2.
5. Woods nickel strike. The bath contained nickel chloride 240 g , hydrochloric acid (S.G. 1.18) 86 cm^3 and made up to 1 dm^{-3} with deionised water.   The spiral was immersed for 1–2 minutes at a temperature of 22°C and electroplated at a current density of 5 A dm^{-2}.
6. Rinsed, (deionised water), washed, (alcohol), dried and weighed.
7. Electroless nickel, (Duraposit 90).
8. Rinsed, (deionised water), washed, (alcohol), dried and re-weighed.
9. Internal stress was calculated from equations {3} and {4} in the Appendix.

3.5 Orthophosphite Concentration

The procedure adopted for the analysis of the sodium orthophosphite concentration followed the instructions in the manufacturer's data sheet for Duraposit 90 *.

3.6 Phosphorus Content Determination Within Electroless Nickel Deposits

Sample preparation: A sample of electroless nickel foil was prepared firstly, by passivating the Pyrene steel substrate in chromic acid (1100 g/dm^{-3}). After thorough rinsing to remove any excess chromic acid, the passivated steel was plated in various aged electroless nickel baths for 1 hour at 92°C. Once plated, the nickel-phosphorus (Ni-P) foil was stripped from the steel, dried and weighed. The phosphorus content of the foil   then determined by ion chromatography.

3.7 Hardness Measurements

The hardness of various aged electroless nickel deposits was analysed using a M41 Vickers micro hardness instrument utilising a diamond pyramid indenter. The samples were etched to distinguish the electroless nickel deposit from the mild steel substrate in 1% nital (1% nitric acid (S.G. 1.42) in ethanol). A 50-100 g load was used and an average Knoop hardness for 10 measurements was taken. The area of interest was viewed under low magnification (× 40) before the pyramid indenter was placed over the deposit. This procedure was repeated for the samples heat treated at 400 °C for 1 hour.

4 RESULTS AND DISCUSSION

Figure 1 shows the results of the stress measurements. Initially, the fresh bath produces a compressive internal stress of 37–24 MPa at 0–1 metal turnovers. The stress decreases with

---

* Duraprep and Duraposit are trade names for chemicals supplied by Shipley Europe Ltd

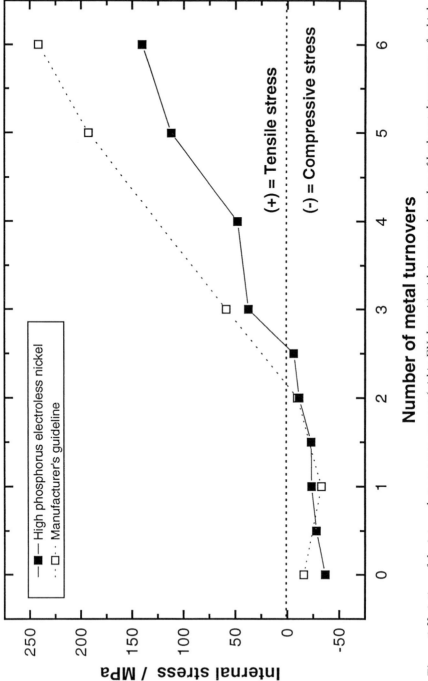

Figure 1 *Variation of the internal stress measurement (within EN deposit) with increased number of bath metal turnovers of a high phosphorous electroless nickel plating solution*

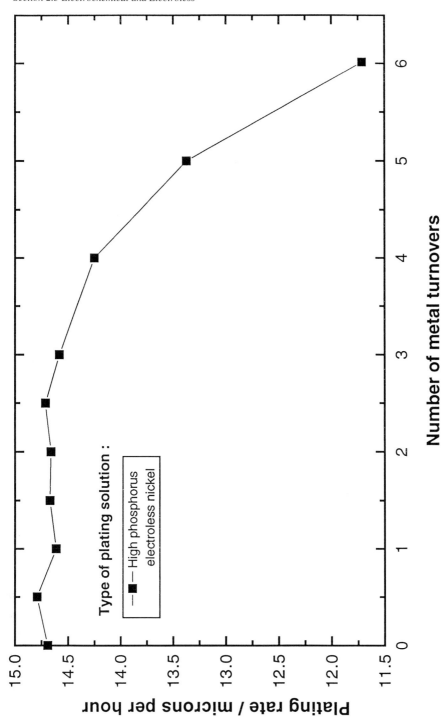

Figure 2 *Plating rate of a high phosphorous electroless nickel solution from baths aged by 0–6 metal turnovers*

increased number of metal turnovers, and after 3 turnovers, the stress becomes tensile. The value of this tensile stress increases as the bath ages reaching a value of 241 MPa after 6 metal turnovers. This result is consistent with other workers[13] and the manufacture's data for the electroless nickel bath used in this work (see Figure 1), but in disagreement with the findings of Bleeks and Brindisi[14] and Keene[15]. The high phosphorus electroless nickel system investigated by Bleeks and Brindisi, showed that the internal stress of the deposits measured by a gearless spiral contractometer, all remained compressive even after 6 metal turnovers. Work carried out by Keene also showed that the internal stress within deposits plated from a high phosphorus electroless nickel bath remained compressive up to 6 metal turnovers.

The plating rate *vs.* the number of bath metal turnovers is illustrated in Figure 2. The plating rate remains fairly consistent until 4.5 metal turnovers when the plating rate suddenly starts to fall. This lowering of the plating rate coincides with the higher values of tensile stress also found after 4–5 metal turnovers and these two factors may be related.

Results of the analysis of the bath for orthophosphite concentration are shown in Figure 3 where there is a linear relationship between the concentration of the orthophosphite ion and the number of bath metal turnovers. Linear regression analysis of the line produced in Figure 3 is illustrated in equation {1} below:

$$c = mn + K \qquad\qquad \{1\}$$

where:

$m = 34.19054$ (sd 0.059652), $K = 0.32112$ (sd 1.88399),
($R = 0.9987$, N = 10 and SD = 3.51515)

During electroless nickel deposition hypophosphite oxidation to orthophosphite is as follows:

$$H_2PO_2^- + H_2O \rightarrow H_2PO_3^- + 2H^+ + 2e^- \qquad\qquad \{2\}$$

As the bath ages, the concentration of the orthophosphite in the plating solution must increase as illustrated by the above equation {2}.

Comparing Figure 3 with Figure 1, it appears that the high levels of tensile stress maybe associated with orthophosphite concentrations in excess of 125 g dm^{-3}. This condition corresponds to a bath metal turnover of approximately 4.5 which is also the point at which the plating rate begins to fall as illustrated in Figure 2.

The phosphorus content of the deposit was found to be in the range of 11.3–13.3 % (11–

Table 1 *Variation of deposit thickness employed in electrochemical tests and plating rate determination*

Plating time/minute	Thickness (μm)
5	1.2
15	3.5
30	7
60	14
90	21
120	28

Table 2 *Variation of the phosphorous content incorporated within the deposit plated from baths aged from between 0–6 metal turnovers.*

Wt. % Phosphorous	Number of metal turnovers
11.3	0
11.9	1
12.1	2
12.0	3
12.2	4
12.4	5
13.3	6

13 % is the figure quoted by the manufacturer). This value was maintained throughout irrespective of the number of metal turnovers. The amount of phosphorus incorporated within the deposit, however, gradually increased with the number of metal turnovers. This is illustrated in the plot of weight % phosphorus vs. number of metal turnovers (Figure 4) and Table 2. This may not be directly related to the age of the solution, but as a result of the plating rate. Generally, factors altering the plating rate also tend to affect the amount of phosphorus incorporated within the deposit. An increase in the deposition rate coincides with a fall in the phosphorus content although the reason for this is still uncertain. As shown in Figure 2 the plating rate falls with increased orthophosphite levels and this may explain the gradual increase of phosphorus observed in the deposits plated from baths with a higher number of metal turnovers.

5 POLARISATION STUDIES

Typical Tafel plots (E *vs.* log i) for various thicknesses of electroless nickel deposits are shown in Figure 5 for a 3 metal turnover bath. From the Tafel plots, values of E_{cor} (corrosion potential) can be determined. These are shown in Table 3 and closely mirror the open circuit potentials obtained in separate experiments.

An uncoated mild steel sample had a E_{cor} of ~ -0.55 V *vs.* S.C.E. while the 28 mm electroless nickel deposit gave a value of -0.27 V. The latter potential may be assumed to be that of a complete electroless nickel coating as no pores or defects have been observed in these samples using accelerated corrosion or chemical porosity tests[16]. From Figure 5, it is apparent that the potential steadily increases in a positive (or noble) direction with increased thickness of deposit. By measuring the corrosion potential (E_{cor}) of various thicknesses of electroless nickel deposits (1.2–28 mm), an assessment of the amount of exposed iron can be made as both electroless nickel and iron have different corrosion potentials. If the surface of the electroless nickel deposit is porous exposing iron at the base of the pores, the E_{cor} value becomes intermediate between the two values of iron and electroless nickel. The more negative the value of E_{cor}, the greater is the percentage of exposed iron.

By extrapolation of the cathodic Tafel slopes, a value of i_{cor} was determined. In 0.125M H_2SO_4 electrolyte, mild steel (i_{cor} = 55 mA cm^{-2}) corrodes faster than electroless nickel

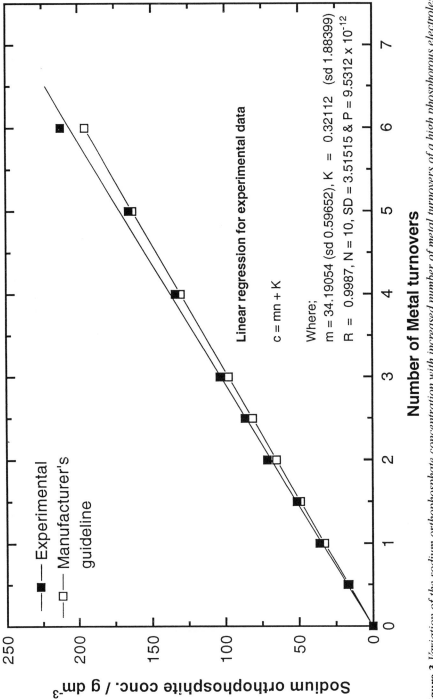

Figure 3 *Variation of the sodium orthophosphate concentration with increased number of metal turnovers of a high phosphorous electroless nickel bath*

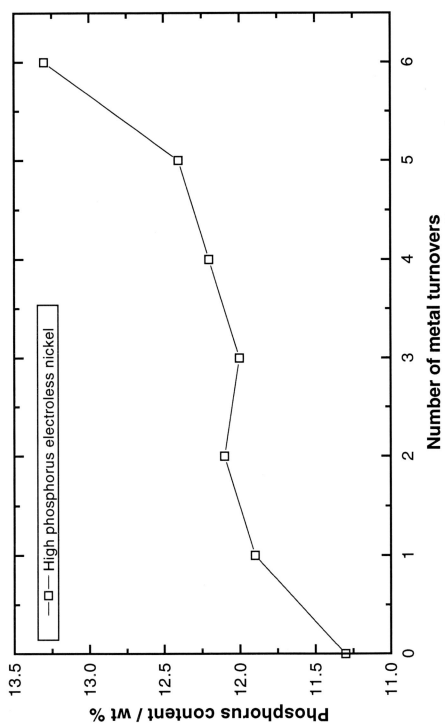

Figure 4 *Phosphorus content/Wt. % of electroless nickel deposit plated from baths aged between 0–6 metal turnovers*

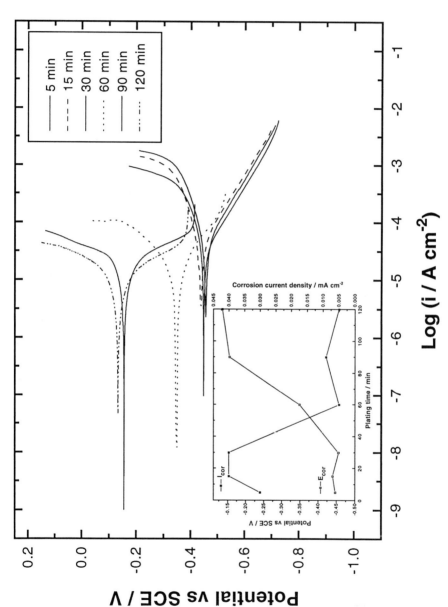

Figure 5 *E-log plots for various EN deposits plated from a bath aged by 3 metal turnovers*

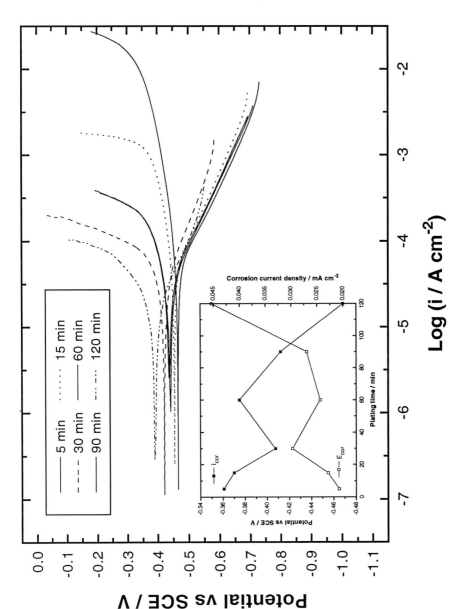

Figure 6 *E-log plots for various EN deposits plated from a bath aged by 4 metal turnovers*

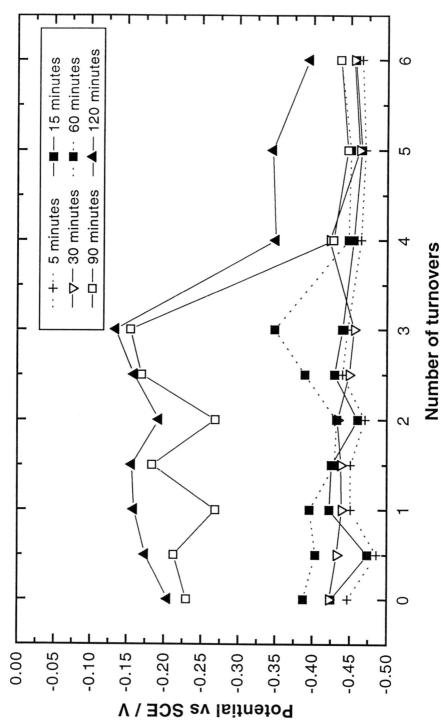

Figure 7 E_{cor} *vs. metal bath turnover for a range of high phosphorus electroless nickel deposits in 0.125M H_2SO_4 at a temperature of 22°C*

Table 3 E_{cor} *and* i_{cor} *data for deposits plated from a bath aged by 3 metal turnovers*

Plating time (min)	E_{cor} vs SCE (V) measured by digital voltmeter	E_{cor} vs SCE (V) measured by potentiostat	i_{cor} (mAcm^{-2}) determined by cathodic Tafel extrapolation
5	– 0.464	– 0.447	0.03
15	– 0.453	– 0.440	0.04
30	– 0.426	– 0.456	0.04
60	– 0.372	– 0.349	0.005
90	– 0.202	– 0.155	0.009
120	– 0.197	– 0.135	0.005

(i_{cor} = 8 mA cm^{-2}) and it is possible to relate a value of i_{cor} to the percentage of iron exposed at the base of the pores. As deposit thickness increases, i_{cor} decreases as illustrated in Table 3 which demonstrates that the pores were filled as the deposit becomes thicker. Virtually identical results were obtained for the freshly made up bath and baths having been aged by 1,2 and 3 metal turnovers. However, there is a noticeable change in Tafel behaviour at 4 metal bath turnovers as shown in Figure 6. It is immediately apparent that E_{cor} values are much closer together for all thicknesses studied and the value of i_{cor} has increased as illustrated in Table 4. Figure 7 shows the variation in E_{cor} with metal bath turnover for a range of electroless nickel deposits. For thicker deposits, there is a tendency for E_{cor} to become more base after 3.5 metal bath turnovers. For thin deposits of less than 19 mm, the trend is not so obvious, with the value of E_{cor} remaining fairly constant ($\sim$ –0.45 V $vs.$ S.C.E.). Figure 8 shows the variation of i_{cor} with metal bath turnover. Once more, there is a marked increase beyond 3.5– 4 metal turnovers. This correlates with the data from plating rate determination, stress measurement, orthophosphite and phosphorus content of the deposit analysis as described earlier.

Anodic polarisation scans for various electroless nickel plated from baths aged between 0– 6 metal bath turnovers are given in Figure 9. There is a change in both the appearance of the E-log i plot and the value of E_{cor} occurring between 3 and 4 metal turnovers. Figure 10 shows the anodic polarisation curve of the electroless nickel deposit after 3 metal turnovers. This

Table 4 E_{cor} *and* i_{cor} *data for deposits plated from a bath aged by 3 metal turnovers*

Plating time (min)	E_{cor} vs SCE (V) measured by digital voltmeter	E_{cor} vs SCE (V) measured by potentiostat	i_{cor} (mAcm^{-2}) determined by cathodic Tafel extrapolation
5	– 0.471	– 0.465	0.043
15	– 0.459	– 0.455	0.041
30	– 0.436	– 0.423	0.033
60	– 0.456	– 0.448	0.04
90	– 0.399	– 0.436	0.037
120	– 0.333	– 0.350	0.01

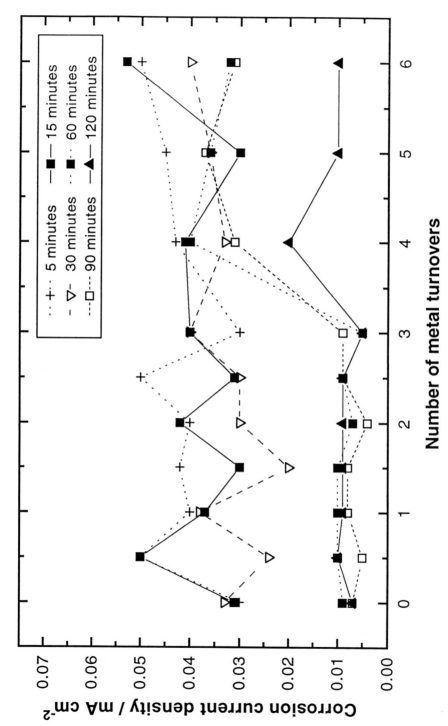

Figure 8 i_{cor} vs. metal bath turnover for a range of high phosphorus electroless nickel deposits in $0.125M\ H_2SO_4$ at a temperature of $22°C$

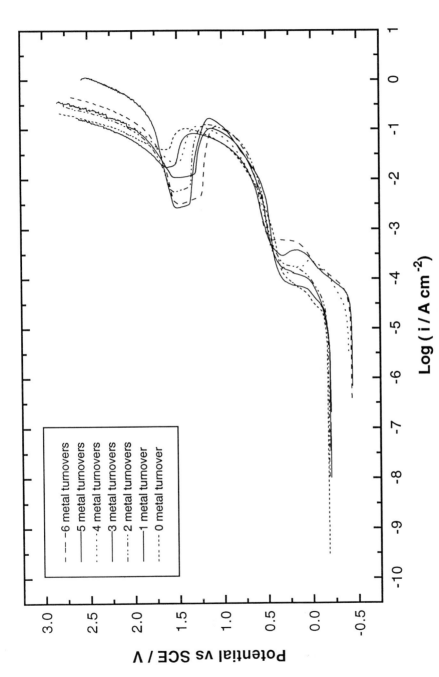

Figure 9 *Anodic polarisation scans of 14 micron electroless nickel deposit plated from solutions of varying age in 0.125M H_2SO_4 at a temperature of 22°C*

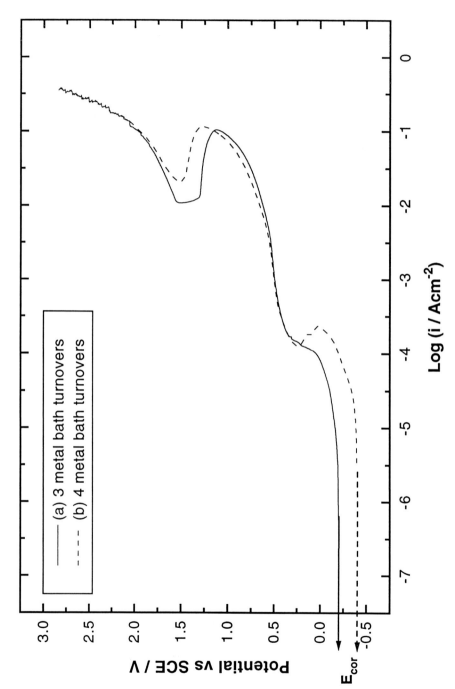

Figure 10 *Anodic scans of a 14 micron deposit plated from an electroless nickel solution aged by 3 and 4 metal turnovers*

result is typical for the deposits plated from 0-3 metal turnovers aged solutions. Between the potential values of 0.84 V and 0.418 V (*vs.* S.C.E.), there is a suppression of the anodic current resulting in a small inflection in the E *vs.* log i curve at a current density value of 0.22 mA cm^{-2}. This suppression of the current is believed to be associated with the phosphorus content of the plated deposit[17]. The corrosion potential for this 14 mm (60 minute) deposit is −0.200 V (*vs.* S.C.E.). However, after 4 metal turnovers, there appears to be a noticeable change and the anodic scan taken from a 14 mm deposit plated from a bath aged by 6 metal turnovers is shown in Figure 10 (b). In the region between −0.126 and 0.392 V (*vs.* S.C.E.), there appears to be a region of anodic current suppression. This occurs at potentials more base than for scans produced for 0-3 metal turnovers, as illustrated in Figures 10 (a) and 10 (b). This region is not stable and does not produce a well defined inflection feature at 0.21 mA cm^{-2} which may be the result of a change in the deposit's phosphorus content. Another significant difference between the deposits plated from baths of increased age is the value of E_{cor}. This value decreases from −0.200 V to −0.390 V (*vs.* S.C.E.) between 3 and 4 metal bath turnovers.

The changes in E_{cor} and appearance in the shape of the regions associated with the suppression of the anodic current between 3–4 bath metal turnovers is consistent with the findings compiled from other experiments during this present work.

6 HARDNESS MEASUREMENTS (KNOOP H_k)

It is clear from the hardness measurements illustrated in Figure 11 that the values of H_k found in the as plated condition (H_k 550–600) remain consistent for a 12-14 mm deposit plated from various electroless nickel solutions (0–6 metal bath turnovers). Heat treated coatings show a marked increase in hardness compared to the as plated condition. Heat treated samples, again, show no noticeable difference in H_k value (H_k 900–1000) between deposits plated from 0–6 metal turnovers. Hardness properties are, therefore, not dependent on factors influencing the ageing process of electroless solutions. Hardness is attributed to the phosphorus content of the deposit, and this present work has illustrated that the phosphorus content within the coating remains fairly constant, whether it is plated from a fresh bath or an aged solution. With the levels of deposit phosphorus remaining consistent irrespective of bath age the level of hardness of these deposits was also found to follow the same trend.

7 CONCLUSIONS

The present research has shown the characterisation of electroless nickel deposits from acid baths alters as the bath ages. The ageing of the bath is directly related to metal turnover additions. Between 4 and 5 metal turnovers, this change is most noticeable as can clearly be seen in Figure 12 with the plating rate decreasing from 14 to 9.5 μm/hr^{-1}. This decline in plating rate is most likely to be attributed to the rise in specific gravity of the plating solution which is primarily caused by the build-up in concentration of sodium orthophosphite, a by-product of the oxidation process. The level of sodium orthophosphite builds up on a linear scale as the bath ages rising to a maximum level of 190 g dm^{-3} after 6 metal turnovers. At this point the solution has reached the end of its useful working life.

The internal stress within the deposit becomes highly tensile between metal turnovers 4 and 5, and this demonstrated in Figure 12. The internal stress measured within the deposit is

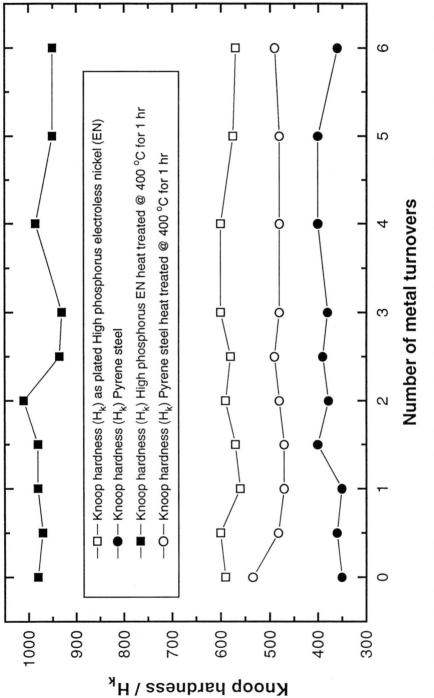

Figure 11 *Knoop hardness (H_k) measurement taken from samples plated from solutions aged by 0–6 metal turnovers before and after heat treatment (1 hour at 400ºC)*

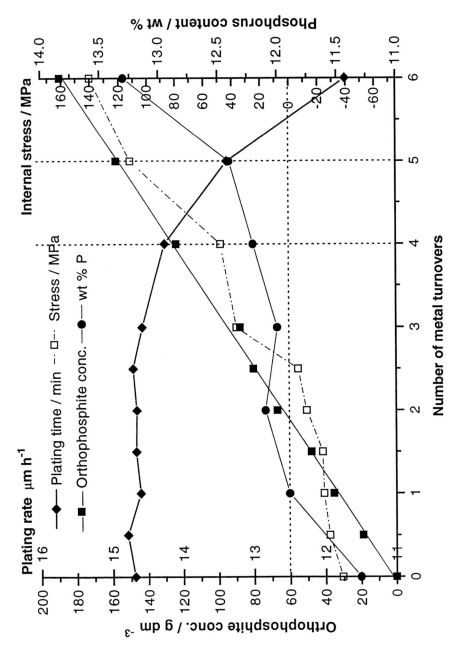

Figure 12 *Various physical parameters vs. number of metal turnovers*

thought to be associated with the amount of phosphorus included in the coating. There are conflicting reports in the literature, however, to the precise nature of the role of phosphorus. The results presented in this paper closely mirror those reported by Riedel and his co-workers, but are in disagreement with those published by Bleeks, Brindisi and Keene. The latter showed that with high phosphorus alloys, the internal stress remains compressive in deposits plated from solutions aged by 0–6 metal turnovers. From analysis of electroless nickel foils, the ageing process had little effect on the weight % phosphorus found with levels rising slightly, with increased number of metal turnovers. From the results reported in this paper and those of other workers, it is clear that the role of phosphorus and the part it plays on the internal stresses within the Ni-P deposit is not well understood.

Corrosion resistance of high phosphorus electroless nickel deposits alters with the number of metal turnovers. Once more, this is most noticeable between 3 and 4 metal turnovers, and can clearly be seen in Figures 5 and 6. Between 3 and 4 metal turnovers both the values of E_{cor} and i_{cor} change significantly, with E_{cor} becoming more base and the corrosion current density becoming higher. The change may be the result of an oxide film covering the surface of the amorphous alloy cracking due to the changes occurring within the deposit caused by the stresses becoming highly tensile. These cracks once formed create local anodic (deposit) and cathodic (oxide film), thereby, increasing the reactivity of these sites. This results in a lowering of the corrosion potential and an increase in the corrosion current density.

This change is again highlighted in the anodic scans of electroless nickel deposits produced from solutions aged by 3–4 metal turnovers (Figure 10 (a) and (b)) in 0.125M H_2SO_4. The anodic current inflection, at 0.22 mA cm^{-2} (Figure 10 (a)), is believed to be associated with the phosphorus content of the coating. The anodic polarisation scan taken from a 14 mm deposit plated from a bath aged by 4 metal turnovers does not produce a current inflection in this region (Figure 10 (b)). However, there does seem to be a disruption in the anodic current in this area. The reasons for this, once again, are unclear.

Appendix

Calculation of the internal stress measured within the deposit

$$\text{Thickness inches} = \frac{W}{AF} \qquad \{3\}$$

where: W = weight of deposit (g), A = area of stainless steel spiral (12 in^2) and F = density of electroless nickel × (2.54)^3.

$A = 12$ in^2 (area of stainless steel spiral)
$F = 2.54^3$ x Density of electroless nickel (7.89 g / cm^3)

$$\text{Stress} = 9.13 \times 10^{-2} \, D/T \qquad \{4\}$$

where: T = thickness of deposit (in) and D = deflection of the spiral contractometer.

(Clockwise (+) denotes a tensile stress and Anti-clockwise (−) denotes a compressive stress)

Conversion of p.s.i. (pounds per square inch) to MPa

Value in p.s.i. $\times$ 6.895 = Value in kPa

Value in kPa $\times$ 10^{-3} = Value in MPa

References

1. B. D. Barker, *Trans. Inst. Metal Fin.*, 1993, **71**, 121.
2. C. Kerr, B.D. Barker and F. C. Walsh, *Trans. Inst. Metal Finishing*. To be published.
3. D. Haywood, *Corros. Coatings S. Africa*, 1992, **19**, 11.
4. H. G. Darkin, *Product Finishing*, Feb, 1990, 6.
5. G. J. Shawhar, *Metal Finishing*, 1988, **86**, 87.
6. H. Deng and P. Moller, Proc. AESF Annual Conf. 79th, 1992, Vol. 2, p. 803.
7. S. S. Tulsi, *Trans. Inst. Metal Fin.*, 1986, **64**, 73.
8. R. Walker, *Metallurgia*, 1968, Oct, 131.
9. C. Baldwin and T.E. Such, *Trans. Inst. Metal Fin.*, 1968, **46**, 73.
10. J. W. Dini, 'Electrodeposition. The material science of coatings and substrates', Noyes Publications Ltd., New Jersey, 1993.
11. J. B. Kushner, 'A new instrument for measuring stress in electrodeposits', 41st Annual Technical Proceedings, American Electroplaters Soc., 1954, p. 188.
12. A. Brenner and S. Senderoff, Proc. American Electroplaters Soc., 1948, Vol. 35, p. 53.
13. W. Riedel, 'Electroless Nickel Plating', Finishing Publications Ltd, Stevenage, England, 1991.
14. T. W. Bleeks and F. Brindisi, Jr., 'The properties and characteristics of electroless nickel coatings applied to gas turbine engine components', Presented at the Gas Turbine and Aeroengine Congress and Exposition – June 4–8, 1989 – Toronto, Ontario, Canada. (The American society of mechanical engineers 89-GT-4).
15. R. H. Keene, 'Application and control of electroless nickel processes at North West Airlines', *Plating and Surface Finishing*, December 1988.
16. C. Kerr, B. D. Barker and F.C. Walsh, 'Electrochemical techniques for the evaluation of porosity and corrosion rate for electroless nickel deposits on steel', Paper for *Trans. Inst. Metal Finishing*. To be published.
17. A. Krolikkowski, "Passive characteristics of amorphous Ni-P alloys', Modification of Passive Films European Federation on Corrosion – Publication 12, 1994, Vol. 12, p. 119.

Contributor Index

This is a combined index for all three volumes. The volume number is given in roman numerals, followed by the page number.

Subject Index

This is a combined index for all three volumes. The volume number is given in roman numerals, followed by the page number.